THE SCIENCE OF
STEPHEN
KING

Books by Lois H. Gresh and Robert Weinberg

The Science of James Bond

The Science of Superheroes

The Science of Supervillains

The Science of Anime

The Termination Node

The Computers of Star Trek

Books by Lois H. Gresh

Chuck Farris and the Tower of Darkness

Chuck Farris and the Labyrinth of Doom

Chuck Farris and the Cosmic Storm

The Truth behind a Series of Unfortunate Events

Dragonball Z

TechnoLife 2020

The Ultimate Unauthorized Eragon Guide

The Fan's Guide to the Spiderwick Chronicles

Exploring Philip Pullman's His Dark Materials

Books by Robert Weinberg

Secrets of X-Men Revealed

A Logical Magician

A Calculated Magic

The Black Lodge

The Dead Man's Kiss

The Devil's Auction

The Armageddon Box

THE SCIENCE OF STEPHEN KING

FROM *CARRIE* TO *CELL*, THE TERRIFYING TRUTH BEHIND THE HORROR MASTER'S FICTION

Lois H. Gresh
Robert Weinberg

BICENTENNIAL
1807
WILEY
2007
BICENTENNIAL

John Wiley & Sons, Inc.

Published by John Wiley & Sons, Inc., Hoboken, New Jersey
Published simultaneously in Canada

Wiley Bicentennial Logo: Richard J. Pacifico

Design and composition by Navta Associates, Inc.

For general information about our other products and services, please contact our Customer Care Department within the United States at (800) 762-2974, outside the United States at (317) 572-3993 or fax (317) 572-4002.

Wiley also publishes its books in a variety of electronic formats. Some content that appears in print may not be available in electronic books. For more information about Wiley products, visit our web site at www.wiley.com.

ISBN 978-0-62045-657-6

10 9 8 7 6 5 4 3 2

To Mark Bocko, Robert McCrory, and Arie Bodek,
who enabled me to become immersed in science.
—Lois H. Gresh

To Dr. Nureen Suwan, Dr. Angela Silva, and
Dr. Carrie Stisser for rescuing my feet!
—Robert Weinberg

CONTENTS

INTRODUCTION

WHERE SCIENCE AND FICTION INTERSECT

I am the literary equivalent of a
Big Mac and fries.

—Stephen King, *Secret Windows: Essays
and Fiction on the Craft of Writing*

Some people are born to be writers. That's certainly true of Stephen King. According to King, as posted on his official Web site, www.stephenking.com, "I was made to write stories and I love to write stories. That's why I do it."

Stephen King started writing while attending grade school in Durham, Maine, and continued to do so throughout his years in Lisbon Falls High School. In 1960, when he was thirteen, he submitted his first story to a magazine; it was rejected. Around the same time, King discovered a box of old science-fiction and horror paperbacks in his aunt's house and was immediately hooked on both genres. He went on to edit the school newspaper, the *Drum*, and also wrote for the local newspaper, the *Lisbon Weekly Enterprise*. He sold his first short story, "The Glass Floor," to Robert A. W. Lowndes's magazine *Startling Mystery Stories* in 1966. It appeared in the fall 1967 issue. Interestingly enough, the same poorly distributed, penny-a-word publication also printed the first short story by F. Paul Wilson, "The Cleansing Machine," in its eighteenth issue, in March 1971. Bob Weinberg, who attended college in Hoboken, New Jersey, from 1964 through 1968, visited the offices of *Startling Mystery Stories* in New York City a number of times during that period and submitted several stories to editor Lowndes. None of them sold.

From 1966 to 1970, King studied English at the University of Maine at Orono, where he wrote a column called "King's Garbage Truck" for the university magazine. He met Tabitha Spruce when they were both working in the campus library. In 1970, King graduated with his bachelor of arts degree in English and obtained a certificate to teach high school. He married Tabitha in 1971 and got a job teaching English at Hampden Academy in Hampden, Maine. During this period, he, his wife, and his two children lived in a trailer.

King wrote horror short stories that he sold to men's magazines such as *Cavalier* to help pay the bills. In spring 1973, he sold his novel *Carrie* to Doubleday Books for a modest advance. Later, on Mother's Day, King was informed by Bill Thompson, his editor at Doubleday, that paperback rights to the book had been sold for $400,000. King received $200,000 as his share of the sale, enabling him to stop teaching and become a full-time writer.

While *Carrie* was an entertaining thriller, King's second book, *Salem's Lot*, established him as a horror writer to watch. *Salem's Lot* was an important book in the history of modern horror fiction. In it, King blended the elements of typical mainstream fiction with the blandishments of horror literature. In so doing, he produced a hybrid, a mainstream horror novel that focused on character development and ordinary people while keeping the monsters offstage and mostly in the background.

After *Salem's Lot*, King wrote *The Shining*, and then, in 1978, came *The Stand*. All of these novels succeeded because they dealt mainly with human beings, not monsters.

King's greatest talent is his ability to mix the horrific with the ordinary. His novels, as well as his short stories, feature normal, everyday people who encounter the bizarre, the unworldly, and the monstrous. It's a marvelous way of engaging the reader. There are no superheroes in King's books, no men in capes or scientists with laser rays. Instead, he populates his stories with the people down the block, the woman at the supermarket, the kids playing basketball on the back lot.

The Stand was the first modern horror epic. It set the standard against which all horror epics published afterward were judged—books such as *Swan Song, They Thirst, Carrion Comfort*, and the entire Left Behind series. *The Stand* showed that horror novels could tell big stories, stories that dealt not merely with the lives and deaths of a few people, of a town, or even of a city, but of the world. And yet the lead characters were still very ordinary people who were forced to deal with the strange and the fantastic that came into their lives.

The Stand redefined horror, broke down the barriers that isolated horror in the genre ghetto, and placed horror solidly among the ranks of modern-day best sellers. Yet, for all its success, the novel was probably the most unoriginal of Stephen King's first four books. The story of civilization destroyed by a plague wasn't a new one. It just had never been described in such chilling detail.

From there, it was just a small step to "The Mist," *Misery*, *Tommyknockers, The Green Mile, Desperation, The Regulators*, and many more. During the last thirty years, Stephen King has become an American institution. He ranks right up there with Grandma, the flag, and apple pie. King is one of the best-selling authors of the twentieth century, if not of all time, simply because he writes about ordinary people, and he does a better job of it than anyone else alive today.

It's this quality that makes stories like "The Mist" into classics. This short novel isn't told by scientists conducting the experiment;

rather, it's told by the people across the lake who are affected by that experiment. As with most of King's stories, there's not even a positive statement that the mist was caused by the scientists. It could have just been some sort of cosmic accident. As in "Trucks," where eighteen-wheelers and pickups take over, whatever brought the machines to life is never really made clear. The only fact that matters is that events are happening, and that's enough—thus, the people involved in those events are forced to act. They rarely have time to sit around and wonder what's going to happen next, what happened before, or even who's dead and who's alive. They're much too busy staying alive themselves. Despite the size of Stephen King's novels, there's barely time for the main characters to take a breath. Figuring out what's what is never a major priority. That's as true in *The Stand*, published in 1978, as it is in *Cell*, published in 2005.

While King is defined as a horror writer, most of his stories are based on science or pseudoscience. In "Trucks," which was made into two movies, *Maximum Overdrive* in 1986 and *Trucks* in 1997, trucks become artificially intelligent and kill humans at a roadside diner. Artificial intelligence, or AI, is the root of many good science-fiction stories, and in those cases, the AI is described in detail and comes across as believable, functioning computer systems. In "Trucks," the reader never learns why the vehicles possess AI, how the AI works, or why the AI is used exclusively to gang up on and then kill humans. That's the difference between King's use of science in his horror stories and the good science-fiction writer's use of science. In science fiction, the emphasis is on the science, and it may benefit mankind rather than destroy everything. In Stephen King's horror, mankind rarely benefits from science.

Horror and science tend to go hand in hand. As early as Mary Shelley's *Frankenstein*, people were terrified that new scientific techniques could create monsters, diseases, and death. In the real world, we're confronted with the knowledge that madmen can kill millions of people by using nuclear, chemical, or biological weapons. Our interest is not so much in how these weapons work; rather, they hold us in grim fascination because they can kill us all.

In each chapter of this book, we examine a scientific field and discuss how it relates to one or more of Stephen King's stories, novels, and films, or a whole mix of them. In several instances, we try to give you a framework of what's been done in the past by other writers on the same subject, and then we show the incredible new science in the story that will someday change your life. We even chip in from time to time with little nuggets of King trivia that, most likely, you would have never discovered otherwise.

Most important, though, we hope you're entertained.

1

FROM PROMS TO CELLS
The Psychic World
of Stephen King

Carrie • Firestarter • The Dead Zone
Hearts in Atlantis • Cell • The Green Mile

> They were silent. There was a kind of haunted avidity about
> them, and that feeling was back in the air, that breathless sense
> of some enormous, spinning power barely held in check.
>
> —*Cell*

It All Started with *Carrie*

Any discussion of Stephen King novels and psychic powers must begin with *Carrie* (Doubleday, 1974), his first published book. It was with *Carrie* that King established himself as the premier novelist of the supernatural, the dark, and the bizarre. Story aside for the moment, the novel itself displayed King's ability to write from unusual perspectives. He gave us slants of Carrie, the main character, from multiple viewpoints, newspaper clippings, and expert testimony. The dormant supernatural gifts of Carrie, a high schooler entering puberty, are brought to full force when she becomes enraged at the tauntings and cruelty of her classmates. King wrote about Carrie as if she truly existed, yet we never peer inside Carrie deeply enough to see accounts of her powerful supernatural forces fully from her perspective. Rather, we see Carrie as if she is drawn by everyone who remembers and analyzes her youth.

Carrie became incredibly famous when the movie by the same name was released in 1976, with Sissy Spacek playing the lead role. She was eerie, petite, and not even close to being as grotesque as the Carrie of the novel. In the book, Carrie is overweight, has pimples, and is washed out and terribly plain. Her mother is a raving psychoreligious, psychosexual lunatic who keeps her poor daughter locked in a closet, forced to pray and read the Bible. Carrie is not allowed to wear pretty dresses, go to parties, or even talk to girls who dare to lounge on their lawns in bathing suits and laugh. Carrie is imprisoned throughout her childhood and her early teenage years by the insane woman who is her mother. By the end of the novel, we realize that Carrie's mother possibly knew of her daughter's latent supernatural powers and, hence, wanted to protect the world from her and from any potential offspring. After all, Carrie's grandmother also had the power to move objects telekinetically and to make "things happen." Why inflict another Carrie-type child upon the world? Carrie alone is capable of destroying an entire town and nearly all of its inhabitants.

Carrie grows up surrounded by votive church candles and a shrine to the martyr St. Sebastian. Rather than listening to rock-and-roll music, Carrie hears her mother wailing and shrieking about religion and sin and how horrible it is to be inherently, innately female. To be born female is an unforgivable sin. Any form of sexual expression, even simply having the body parts, is enough to condemn a young girl to hell forever.

Many people are bullied as teenagers or, at minimum, feel like outsiders. They are too fat or too ugly or too dumb. Or maybe they just aren't popular enough to feel comfortable inside their own skin. With the power to destroy their tormentors, would they do it?

Carrie holds off for as long as possible. She endures more than most of us are forced to endure as teenagers. She lives our worst nightmares, and then some. It is to her credit that she doesn't use her supernatural powers until she reaches her breaking point. But when Carrie breaks, so does the world. And here, we see Stephen King at his best: when someone with the power to destroy the world breaks.

So how does Carrie reach a breaking point and unleash her powers? And just what are her powers, and how possible, in reality, are they?

Teenage Sue Snell feels guilty about making fun of and torturing Carrie along with all of the other kids at school. She is essentially kind and good-hearted, and she wants to make things better for poor Carrie. So Sue asks her boyfriend, Tommy Ross, to take Carrie to the senior prom. Carrie is tentative at first, thinking it all a big joke when Tommy asks her to accompany him, but she can't resist an offer to fit in and be part of the crowd, to finally be treated as if she matters just a little. So Carrie, thrilled at her great fortune, accepts Tommy's invitation. Of course, she must battle her mother just to wear a dress or go to the dance, but she persists, thinking it may be her one chance to lead a normal life, if only for one night.

Things do not go well, to put it mildly.

Chris Hargenson, one of the most popular and beautiful girls in the school, and one of Carrie's main tormentors, arranges for a bucket of pig's blood to drop on Carrie the very moment Carrie is falsely crowned prom queen. As Carrie stands on the stage, smiling and happy with the crown perched on her head, the bucket looms above her, ready to drop in a moment. And so it does drop, drenching Carrie in blood.

Carrie snaps. She telekinetically locks all the doors leading from the school. She uses her mind to turn on killer sprinkler systems, as well as to launch killer fires. She destroys everything and anything, yet never moves an inch from the stage. Her mind does all the work for her.

With nearly everyone dead and the high school in flames, Carrie finally walks home and washes the pig's blood from her face, neck, arms, and body. Her mother greets her with, "Thou shalt not suffer a witch to live," and then lunges with a knife in hopes of killing her evil, sinful daughter. Carrie crucifies her mother with kitchen utensils, then destroys the house.

Carrie's supernatural powers are no longer dormant. She has learned to use and control her powers to get what she wants and, most important, to get people to leave her alone. In this way, *Carrie* is similar to the 1973 film *The Exorcist*, in which a young girl's body is possessed by a force she cannot understand or control. Carrie is also possessed by powers she cannot understand or control. In both *The Exorcist* and *Carrie*, the young girl is both a victim and a monster.

The film *Carrie* was a huge hit when it was released in 1976. The sale of the paperback rights for the book enabled Stephen King to give up his previous jobs and concentrate completely on writing. It was one of those rare instances where one book hitting it big changed the author's entire life. Its success spawned a horde of imitators, movies with similar plots but without the same emotional punch that King's book provided. Still, *Carrie* relied heavily on that which had come before. It was not the first book, fiction or nonfiction, that dealt with psi-talents, nor would it be the last. In the next few sections, we'll take a brief but in-depth look at some of *Carrie*'s ancestors and see how they might have provided inspiration for King's masterwork.

Carrie's Ancestors in Fact

Parapsychology is the study of the evidence of mental awareness or influence of external objects without interaction from known physical means. Most objects of study fall within the realm of "mind-to-mind" influence (such as extrasensory perception and telepathy), "mind-to-environment" influence (such as psychokinesis), and "environment-to-mind" activities (such as hauntings). Collectively, these three categories of abilities are often referred to as psionics.

Needless to say, the scientific validity of parapsychology research is a matter of frequent dispute and criticism. Among scientists, such fields are known as pseudoscience, which, by definition, has been refuted by numerous rigorous scientific studies.

Anecdotal reports of psychic phenomena have appeared in every culture since the dawn of history up to the present day. Historically, the existence of such phenomena was commonly accepted even among the highly intelligent. Many early scientists expressed interest in such phenomena.

With the advent of the scientific revolution in the beginning of the nineteenth century, and led by the British Royal Society, a distinction came to be made between "natural philosophers" (later to be termed *scientists* in 1834) and other philosophers. Many of the natural philosophers, including Newton, believed in various types of Renaissance magic, such as alchemy.

Following the scientific revolution was a period known as the Enlightenment. This movement advocated rationality as a means to establish an authoritative system of ethics, aesthetics, and knowledge. During this period, it was proposed for the first time that life should be ruled by reason as opposed to dogma or tradition. The basic view of the world was that the universe worked as a mechanistic, deterministic system that could be studied until everything about it was known through calculation, reason, and observation. Because of this belief, the existence or the activity of deities or supernatural agents was ignored. The Enlightenment was the beginning of the verbal war between people who believed in psychic phenomena and those who thought it was all nonsense.

Mesmerism

Franz Anton Mesmer (1734–1815), a Viennese physician, considered himself a man of the Enlightenment. At the time, electricity and magnetism were thought of as invisible "fluids." Mesmer believed that he had discovered a third type of natural fluid, which he called animal magnetism. Mesmer believed that animal magnetism, if used properly, could heal various ailments without healers resorting to the supernatural. He developed a technique that he termed *mesmerism*. This technique, which produced an altered state of mind, we now call hypnotism. One important discovery that Mesmer made was that during mesmerism, a few people exhibited what he called "higher phenomena," such as apparent clairvoyance.

The most famous person who exhibited this talent was the psychic Edgar Cayce, who entered trances in which he could communicate "in his mind" with an individual anywhere in the world.

Mesmerism never caught on with scientists. In 1784, two studies were authorized by the French Royal Society of Medicine and the French Academy of Sciences. Both groups investigated mesmerism and issued negative reports. One scientist who was associated with mesmerism, Baron Carl Reichenbach (he became famous for his discovery of paraffin fuels), developed a vitalist theory of the Odic force (a vital spark or "soul" radiating from all life) to explain parapsychological phenomena.

Mediums

By the 1850s, mesmerism had run its course. Due in part to changing attitudes in religion, however, the feats of people in mesmeric trances continued to generate a lot of attention. Mesmerism became the foundation for mediums of the newly started spiritualist movement, whose followers claimed to contact the spirits of the dead. By the end of the era, mediums were flourishing throughout all the major cities of Europe.

Unfortunately, most accounts of mediums and psychics left much to be desired. Take, for example, the case of Daniel Dunglas Home, one of the most famous physical mediums of the nineteenth century.

According to popular accounts of Home's life, he was born in Edinburgh, Scotland, in 1833. In 1842, he moved with his family to the United States. As a teenager living with an aunt, he discovered that he possessed the gift to move furniture in the house with just his mind. After being reassured by ministers that his mental powers were a gift to be shared, Home began holding séances. According to people who attended these events, knocking noises were heard, furniture levitated, and an accordion locked in a cabinet played by itself.

In 1852, Home became famous for levitating himself at his séances. In 1855, he returned to England and later toured Europe and Russia. Reports at the time say that he was able to levitate tables so high in the air that he was able to walk under them. Home was accused by many skeptics of being a fraud, but according to newspaper stories of the time, no one ever proved that his psychic gifts weren't real. The same stories reported that while many scientists claimed that Home was merely an excellent stage illusionist, no one could demonstrate how he performed his psychic feats. He died from tuberculosis in 1886.

Home's incredible feats, if true, would classify him as a real-life Carrie. But, as pointed out by the psychic debunker James Randi, Home's claims were merely embellished lies told to a naive public and newspapers looking for sensational headlines. As a teenager, Home was thrown out of school for "demonstrating poltergeist activity" to his fellow students. His séances were performed only to believing audiences, and, despite reports to the contrary, he was exposed as faking illusions numerous times. Most likely, the mysterious accordion music was in reality the sound of a harmonica hidden in Home's mouth.[1]

Fraudulent psychics like Home flourished in Europe and the United States during the nineteenth century. Faced with a world swiftly being explained by science, many religious people clung to any supernatural aspect of life that seemed possible—until science invaded the psychic world.

Using Science to Validate Psychic Phenomena

The basic concept for a learned, scientific society to study psychic phenomena began with the spiritualist E. Dawson Rogers, who hoped to gain a new kind of respectability for spiritualism. The Society for Psychical Research (SPR) was founded in London in 1882. By 1887, eight members of the British Royal Society served on its council. Many spiritualists left the SPR soon after its founding, however, due to differing priorities and skeptical attitudes within the SPR toward mediums. Nevertheless, the SPR continued to research psychic events, publishing its findings in its yearly proceedings. Similar societies were soon started in most other countries in Europe, as well as the American SPR in the United States. The British SPR remained the most respected and skeptical of all these societies.

Most early SPR research involved testing famous mediums and others who claimed to control psychic "gifts." The society also performed some experiments involving cards and dice. Still, the field of psychic research did not gain a measure of respectability until R. A. Fisher and other scientists developed statistical methods for studying psychic events. It was around that time that the name *parapsychology* replaced *psychic research*.

Perhaps the best-known parapsychology experiments were conducted by J. B. Rhine and his associates at Duke University, beginning in 1927. Rhine used the distinctive ESP cards of Karl Zener. The Rhine tests involved much more systematic experiments than those conducted by the SPR. Also, Rhine used average, ordinary people instead of mediums or people who claimed to have gifted abilities, and he used statistical methods to analyze his results.

Rhine published the story of his experiments and their results in his book *Extra Sensory Perception* (1934), which popularized the term ESP. Equally influential was Rhine's second book on the Duke experiments, *New Frontiers of the Mind* (1937). Rhine helped to form the first long-term university laboratory devoted to parapsychology in the Duke University Laboratory. It later became the independent Rhine Research Center. He also helped to found the *Journal of Parapsychology* in 1937, which remains one of the most respected journals in the field today.

Inspired at least in part by Rhine's experiments, the U.S. government has conducted a number of investigations into parapsychology. Perhaps the most famous of these is Project Stargate, conducted by the CIA and the Defense Department in the late 1970s and early 1980s. These experiments involved the ESP talent known as "remote viewing," where the test subject can telepathically see a scene from hundreds of miles away. As of yet, none of these government projects has yielded any significant results— at least, none that have ever been made public. We introduce the idea of using humans with psi powers as secret government agents later in this chapter when discussing Stephen King's novel *Firestarter*.

Carrie's Ancestors in Fiction

The belief that some people are gifted with more than five senses is thousands of years old. Long before there were books, there were tales and legends of people with "second sight," or the ability to tell the future. Perhaps the first such story was that of Cassandra, from the *Iliad* and the *Odyssey*. Cassandra is cursed by Apollo to prophesy the truth but never be believed. Thus, she warns her father, Priam, the king of Troy, that the city will be conquered by Greeks, but he doesn't listen to her. Shakespeare used prophecies in his plays: Caesar is warned to "beware the Ides of March," while Macbeth is hailed by the third witch with, "Thou shalt be king hereafter."

Lost Races and Superhumans

The inspiration for *Carrie* and the entire genre of psychic-powers-run-amok novels that followed it came from the thriller and horror novels of the late nineteenth century. Inspired at least in part by the famous mediums and the formation of the Society for Psychical Research in 1882, these novels often featured women in peril who found their salvation through psychic means. Also extremely popular were lost-race novels describing hidden tribes that possessed mental powers far beyond those of ordinary men. Some of the most famous lost-race novels of the period portrayed characters with ESP powers. These books include *She* by H. Rider Haggard (1887), *Thyra—of the Polar Pit* by Robert Ames Bennett (1900), and *Eric of the Strong Heart* by Victor Rousseau (1914). Sax Rohmer's tales of evil Asian villains often depicted good and evil characters with psychic powers, beginning with the most famous of all "yellow peril" novels, *The Insidious Dr. Fu Manchu* (1914). Olaf Stapledon's *The Last and First Men* (1930) describes the future of mankind for the next several billion years and includes numerous mutations of humanity into beings with tremendous psychic powers.

Stapledon further explored the theme of superhumans in his 1936 novel *Odd John*. The book deals with a mutant who has extraordinary mental powers, including telepathy, from his earliest youth to his death at age twenty-three. While hailed by many critics as a classic, *Odd John* portrays its lead character as an amoral and selfish superman. In one episode, John kills a policeman who catches him stealing; he reasons that normal humans are so far beneath him that killing them is no more a crime than is executing a dumb animal. Published during the Nazi rise to power in the 1930s, *Odd John* served as a grim statement of the danger of power without morality.

Science Fiction and ESP

It was in the science-fiction magazines of the 1930s and 1940s that ESP powers got their biggest boost. In numerous interviews, Stephen King has acknowledged his debt to the science-fiction and fantasy publications of that period.[2] Early stories such as Edmond Hamilton's "The Man Who Saw the Future" (1930) and "The Mind-Master" (1930) are cautionary tales that warn of the perils of psychic powers. By the late 1930s, though, readers had grown tired of stories that dealt with the dangers of scientific research. In September 1937, E. E. Smith's serial novel *Galactic Patrol* introduced the character Kimball Kinneson of the Lensmen, a legion of telepathic guardsmen who fight for the peace of the galaxy. Kinneson is only one of a number of heroes using ESP powers against evil space overlords known as Boskone, who are out to rule the universe. Kinneson starred in four novels over the next ten years—*Galactic Patrol*, *Gray Lensman*, *Second Stage Lensman*, and *Children of the Lens*. He is aided by telepathic members of dozens of alien races in his efforts to defeat the Boskone in a cosmic battle much bigger than he ever imagined.

In 1940, science-fiction fans hailed *Slan* by A. E. van Vogt, serialized in *Astounding Stories* from September to December, as the best story of the year and one of the best of the previous decade. Van Vogt wrote *Slan* partly because the editor of *Astounding Stories*, John W. Campbell, said that it would be impossible for a normal person to write a novel about a superhuman being. After buying *Slan* for the magazine, Campbell admitted that van Vogt had proved him wrong.

Slan tells the story of mutants, called slans, living on Earth in the future, who have the power to read minds. Normal people, fearing that the slans are the next step in evolution, hate the mutants and try to kill them all. *Slan* describes how one young mutant supergenius, Jommy Cross, discovers a way to end the human-mutant war. One of the most popular science-fiction novels of the 1940s, *Slan* helped to popularize stories of mutants with superhuman mental powers.

Equally entertaining, though much darker than *Slan*, is Norvell Page's short novel *But without Horns* (*Unknown*, June 1940). Page took Campbell's claim that no one could write a believable novel about a superman and turned it upside down. The story of a man with such strong telepathic powers he can project illusions and take control of men's minds, *But without Horns* is a superman novel in which the superman never appears. The villain of the novel remains offstage during the entire novel. It is the ultimate paranoia novel, and Campbell was once again forced to admit he was wrong about superman novels.

Written in a much more subtle and subdued style was the In Hiding series, authored by the schoolteacher Wilmar Shiras and published in *Astounding SF* from 1948 through 1951. The stories, dealing with children who have superhuman intelligence and ESP powers and their attempts to remain in hiding among ordinary students, were collected in book form as *Children of the Atom* (1954). *Children of the Atom* was one of the most influential science-fiction novels ever published, as it was read by thousands of baby boomers who felt a strong kinship with the heroes and the heroines of the novel. It was the book that helped to make science fiction the literature of the 1980s onward.

The 1950s, with its worries about the atomic bomb, fallout, and mutant children, saw a rise in the number of stories dealing with ESP powers. Among the best were the novels *The Power* (1956) by Frank Robinson, *Highways in Hiding* (1955) by George O. Smith, and *Jack of Eagles* by James Blish (1952). One of the most interesting novels of the period was *The Demolished Man* (Shasta, 1953) by

Alfred Bester, in which all the people of the future possess the power of telepathy and one man schemes to commit murder. Bester came back a few years later to top his first book with *The Stars My Destination*, where future society is centered around the ability to teleport. Many science-fiction critics considered the second Bester novel the top science-fiction novel ever written.

Another fascinating novel based on ESP powers was Philip K. Dick's *The World Jones Made* (1956). In the story, Jones possesses the ability to see into the future one entire year. He is thus able to engineer his takeover of the world. But there is no joy in Jones's life because soon after he accomplishes his mission, he foresees his own death taking place a year later. No matter what he does, he always knows that he will die.

One of the funniest series of science-fiction novels ever written were the three Mark Phillips novels dealing with ESP powers, published in *Astounding SF* from 1959 to 1961. Phillips was the pen name for Randall Garrett and Laurence M. Janifer. The trio of novels depict an FBI agent named Malone who is assigned to capture crooks who use psi powers to commit crimes. In *That Sweet Little Old Lady*, Malone discovers that a foreign spy is reading the minds of important government officials, so he sets out to find a telepath of his own to discover the mind reader. Unfortunately, the only telepath he can locate is a loony old lady who thinks she is Queen Elizabeth I.

In the second novel of the series, *Out Like a Light*, Agent Malone is sent out to find an auto-theft ring that steals cars by teleporting them away. How to stop a criminal who can transport his body somewhere else just as you are about to put handcuffs on his wrists is another challenge for Malone's cunning. A third novel, *Occasion for Disaster*, has Malone trying to prevent a global disaster brought about by rogue ESP users.

Another writer who specialized in stories dealing with ESP powers was Theodore Sturgeon. His classic novel *More Than Human* (1953) describes how three lonely, isolated people meld their minds to form a single powerful telepathic entity.

John Wyndham's classic science-fiction horror novel *The Midwich Cuckoos* is a masterful tale of a weird alien invasion, which was made into an effective film called *Village of the Damned* (first in 1960, then in 1995). In the story, women at one particular village are struck down by a mysterious signal from outer space. When they recover, they discover that they are all pregnant. When the children are born, it turns out that they possess telepathic powers and seem to have been born to fulfill a certain mission on Earth. The weirdness of telepathic children thinking the same thoughts and controlling other people in the village is handled effectively in the film. The climactic scene, in which the hero has smuggled a bomb into the village to kill the children and the children attempt to break down his mental barriers before the bomb explodes, makes for gripping suspense.

Carrie Battles the Bullies

With *Carrie*, Stephen King took a conventional story of teen alienation and despair and turned it into a shocking horror epic. There were many reasons *Carrie* was a success as a movie, but near the top of any list of explanations has to be "Payback is a bitch." In no uncertain terms, *Carrie* proved this, as did all the other novels and movies that followed in the same footsteps.

Defining Carrie's Powers

The word *psychokinesis* is the more modern term for what is generally known as *telekinesis*. The two words have similar meanings, with *psychokinesis* being defined as "mind movement" and *telekinesis* defined as "distant movement." Whether we call the ability psychokinesis or telekinesis, we're basically talking about what Carrie does: move objects using her mind. She has what is known as a "psi" ability to alter matter simply by thinking that she wants to alter it.

Psi has to do with parapsychology, or the study of mental abilities that enable people to influence objects and people without their physically doing or saying anything. The psi form of mind-to-mind influence includes mental telepathy, which we'll talk about later in this chapter, and ESP. The psi form of mind-to-matter influence includes psychokinesis or telekinesis.

This book focuses on the supernatural science of Stephen King. We can't think of psi, also known as psionics, as an actual science. There's no proof that psychokinesis or ESP exists, for example. Most of what Stephen King writes about is pseudoscience, which refers to knowledge and practices that are incorrectly deemed science. It's not to say that a pseudoscience may not someday become a science. Rather, nothing has been proven conclusively yet, and, hence, a pseudoscience cannot be taken seriously as a field of science.

Of course, parapsychologists, who study phenomena such as Carrie's supernatural abilities, believe that there is evidence of psionic power. Jessica Utts, a statistics professor at the University of California-Davis, works extensively in parapsychology. She claims that psychic power has indeed been proven by statistical studies. She also claims that there are no methodological flaws in the studies. Dr. Utts teaches courses such as "Integrated Studies: Testing Psychic Claims." On her collegiate Web site (http://anson.ucdavis.edu/~utts/), she lists links to parapsychology research papers and labs. One such link (www.parapsych.org/faq_file1.html)—to an FAQ about parapsychology—defines the "basic parapsychological phenomena," among them:

- ESP—obtaining information about events beyond the reach of the normal senses; includes mental telepathy, clairvoyance, and precognition
- Mental telepathy—communication between two minds
- Clairvoyance—obtaining information about events in remote locations, well beyond the reach of the normal senses
- Precognition—knowing the future, having premonitions; includes having dreams about things that haven't happened yet

There are many other types of basic parapsychological phenomena, such as reincarnation, hauntings, and near-death experiences.

The parapsychology FAQ points out that scientists traditionally think in materialistic ways, considering human consciousness to be the intertwined physical functioning of the nervous system, the brain, and the body. They believe that mental functioning cannot possibly include foretelling the future, moving objects, and so forth—it is physically impossible. Yet psychic phenomena have been reported in every culture throughout all of known history. Hence, concludes the FAQ, there is something more to human consciousness than the physical connections among the nervous system, the body, and the brain. In short, we have souls. And, possibly, these souls are responsible for the sort of power possessed by Carrie.

While the FAQ does not delve into the notions of collective consciousness among all human beings and why our souls can give someone like Carrie extraordinary powers, other great thinkers throughout history have explored these ideas. If somehow we are all linked in a cosmic way through our consciousness or souls, then possibly we can manipulate objects, transfer thoughts, and make things happen at far distances. Clearly, these ideas remain unproven, but let's look at the broad strokes behind them to see whether they might have merit.

Consciousness and the Soul: Philosophers Weigh In

It was René Descartes who first wrote the famous line "I think, therefore I am." His immortal statement had to do with the fact that people consider themselves to be separate entities from one another, distinct individuals. We think of ourselves as being individual selves with individual souls. Somewhere in the mix are our minds. Is the soul part of the mind? Is the ability to be a conscious entity, or a self, also part of the mind? Philosophers, scientists, and others have long pondered what these terms mean: *self*, *soul*, and *mind*.

Plato claimed that the gods put our souls into our bodies. These souls were "of another nature,"[3] and inside the digestive system was "the part of the soul which desires meats and drinks and the other things of which it has need by reason of the bodily nature."[4]

According to Aristotle, all objects consist of matter, and as the matter changes, the objects change as well. If, for example, a garment is made of cloth, and you rip the cloth into shreds, then the garment no longer looks the same. If you bake a cake and put icing on the top, then scrape off the icing, the cake will have a different form.

Aristotle took this notion one step further. He considered the soul to be an object, or a form made up of the matter from which a person is created. If a person's matter changes, so does his or her soul. If a person gets sick, the soul may suffer. As a person grows older, so does the soul.

To Aristotle, the soul is the person's caretaker. It handles everything that we require to remain alive. Animals, as well as humans, have souls, and each type of animal has a different type of soul that is suited to its particular animal nature. A different set of conditions from those of a human is required to keep a tiger alive, for example. Aristotle felt that only the human soul has reason and will. Animal souls are not rational, he claimed, and do not have foresight and consciousness. This idea has been debated for centuries. We would argue, for example, that a dog that loves its master and protects him has a soul and consciousness. Is consciousness the same as the self and the mind?

To look for the souls and analyze them, Aristotle dissected many animals, though he didn't go so far as to dissect humans. He guessed that souls live inside our hearts. These souls affect our entire bodies. Yet when he dissected animals, he could not find the souls inside their hearts.

During Aristotle's time, people basically knew nothing about the nervous system. Aristotle believed that our eyes and ears are connected to blood vessels that, in turn, are connected to the heart, where the soul lives. He did not guess that the eyes and the ears are connected to the brain, which is part of the nervous system. Because blood vessels are, of course, connected to the heart, he figured that the heart controls our senses, as well as all movement.

The nervous system was discovered in 322 B.C. by the Greek

anatomists Herophilus and Erasistratus, who dissected human corpses. They discovered that in addition to the system controlled by the heart, people have another system inside their bodies, one that controls what happens in the spine and the skull.

Picking up the theories about the soul from where Aristotle left off, Herophilus and Erasistratus merged their new information about the nervous system into the medical knowledge of the time and came up with their own hypotheses about the human soul. It was their guess that the soul entered the body every time a person took a breath. Somehow, the soul existed outside the person, in a cosmic "everywhere" form, and a person could literally breathe his or her own soul into the body—repeatedly.

If indeed people believe this idea, then it makes sense that they might also conclude that all of our souls are outside our bodies in this cosmic everywhere form, and that we can transfer thoughts and ideas, and we can move things at great distances, simply by willing it to be so. Let's assume that my mind is connected to my soul and that my soul is floating around next to everyone else's soul and that all of these souls are interconnected somehow. Given all these conditions, I should be able to have my soul or consciousness connect with all the others and impart information to, say, your soul or consciousness. This would be called mental telepathy. We see evidence of mental telepathy in King's *The Dead Zone*, which we talk about later in this chapter, and in his recent book *Cell*, in which people are able to communicate telepathically after being infected by their cell phones.

According to Herophilus and Erasistratus, after the soul enters a person's body it flows into the heart. From the heart, the soul flows throughout the body using the arteries and the veins. Along the way, of course, the soul flows into the brain.

Oddly, Herophilus and Erasistratus guessed (rather incorrectly) that the mind is inside the heart. They didn't guess that the mind is inside the brain. So they incorrectly assumed that the mind flows from the heart to the muscles and the brain, and the brain itself has no control over the body.

Galen's Theories of Consciousness

Medical research always progresses, and four hundred years later, in A.D. 150, a doctor named Galen decided to continue the research started by Herophilus and Erasistratus. Galen went to Alexandria, where he served as the medical doctor to gladiators and emperors. While healing these elite patients, Galen also dissected animals every day, and over time he combined all the soul, mind, and consciousness ideas postulated by Aristotle, Plato, Herophilus, Erasistratus, and Hippocrates into new theories.[5]

Galen deduced that each organ in the body has a special role and purifies substances for use by other organs. For example, he said, the stomach attracts all of the food we eat, and it turns the food into chyle. This chyle flows into the intestines and the liver. Galen thought that the liver then transforms the chyle into blood, which flows into the heart.

Inside the liver, after the chyle is changed into blood, the lungs suck all the impurities out of the new blood. From the liver, the purified blood flows into the veins en route to the heart. Inside the heart, the blood mixes with air from the lungs. The air contains vital spirits aka the soul and the collective consciousness. Once again, we see that the ancients supplied medical theories that backed up the notions of mental telepathy and other paranormal capabilities.

Continuing Galen's hypothesis, the blood containing the vital spirits courses throughout the body, including the brain. Before entering the brain, right at the base of the skull, the blood is purified again and turns into what Galen called animal spirits.

These animal spirits are responsible for our thoughts and sensory experiences, and they are also the force that enables us to move our bodies. The vital spirits might suck the cosmic consciousness and raw souls into our bodies, but the resulting animal spirits are the actual individual souls.

Galen taught that the brain is a pump at the top of the body. The mind itself is not part of the brain but, rather, is stuffed into the skull in the empty spaces. The mind, in its cosmic consciousness form, also exists in outer space: in the sun, the moon, and the stars, all of which possess vast intelligence far beyond any intelligence known to man.

Clearly, with such a vast intelligence linking us all together in the way that Galen believed, mental telepathy and other paranormal abilities become more than vague notions. They become possible—if you believe what Galen taught.

Early Christians and Visionary Mavericks

After Galen's time, Christianity eventually absorbed many of his medical theories. The early Christians were particularly interested in Galen's ideas about the mind, the soul, and the collective cosmic consciousness.[6]

In the Old Testament of the bible, the soul is a living entity in our blood, preferring to dwell most often in our hearts and livers. When people die, their souls die with them.

In the New Testament, the soul is immortal and invisible. It lives nowhere and everywhere at the same time, and it imparts special capabilities and feelings inside the skull. When a person dies, his or her soul goes either to heaven or to hell. Either way, it's still hanging around in the vast cosmic consciousness, and by tapping into it, knowledge could be transferred, clairvoyance and speaking to the dead could take place, and other forms of paranormal experiences could happen.

The church, however, didn't believe everything taught by Galen. It still believed some of the things that Aristotle taught, such as the idea that the soul lives inside the heart. So while the church believed in an immortal, invisible soul that is everywhere and nowhere at the same time, with special focus in the skull, it also believed that the soul resides chiefly inside one's heart. It's not incredibly logical to believe all of these things simultaneously, but many medical discoveries were unknown at that time and people made the best guesses they could, based on what little information they had.

It wasn't until much later, for example, that atoms were discovered. With this new discovery came new theories about the soul and the cosmic consciousness.

Epicurus

The Greek philosopher Epicurus suggested that the world consists of invisible particles that control nearly everything. These particles are of the utmost importance. Epicurus suggested further that the gods of his world probably didn't care about human souls at all. Their only concern was with the particles.

As for the soul, Epicurus postulated that it consists of atoms inside the chest. These atoms are constantly seeping from our bodies. As we breathe, we suck the atoms of our souls back inside ourselves, hence restoring our mental and physical balance. So you see, Epicurus was another believer in cosmic consciousness, the very thing that could give Carrie her powers.

According to Epicurus, the soul is exactly that: a cosmic property. Remember that the soul atoms continually seep from the body. When people can no longer breathe enough soul back into their bodies to replenish the soul atoms that they've lost, they die.

Of course, religious people back then did not agree—at all—that when a man dies, his soul dies with him. After all, the church taught that the soul is immortal and either goes to heaven or hell. Famous for his depictions of hell, Dante went so far as to put Epicurus into hell.

Thomas Aquinas

In the thirteenth century, Thomas Aquinas delved into all of the ideas about souls and cosmic consciousness that came before him. Being a theologian and a very smart man, he dismissed Epicurus's claim that we're all part of a cosmic soul made up of the atoms that seep from our bodies. He dismissed the idea that when we die, our souls die with us.

Instead, Thomas Aquinas supported the theory that all good souls go to heaven, and he added that the stars are a twinkling of this beautiful heavenly place. Aquinas agreed with Aristotle that the soul lives in the heart. He agreed that the soul is the form of the matter that constitutes our bodies. It is the form of all life.

Other than simply reverting to what Aristotle taught long ago, Aquinas pushed things a little further. He suggested, for example, that the soul's facilities are in the skull. And while he believed that animals have souls, he also believed that only human souls survive death and live forever in either an immortal bliss or hell.

Aquinas set the stage for natural philosophy in universities. His ideas weren't medical; rather, they were rooted in philosophy and religion.

In the meantime, medical doctors continued to teach Galen's anatomy, dissecting humans and animals to learn more about the body, the brain, the mind, and the soul. Doctors taught medical students that the soul resides in the heart, the liver, and the skull. They also taught that invisible spirits, in the form of the immortal soul, are constantly traversing the entire human body along visible pathways—arteries, the nervous system, and the digestive system.

Andreas Vesalius

At the University of Padua in 1537, the head of surgery and anatomy was a man named Andreas Vesalius. He regularly had students dissect human cadavers, and he started making detailed charts of the human body. He began to think that Galen's medical theories were based on animal dissections rather than on human dissections. He analyzed all that had come before and rethought—yet once again—the ideas about the brain, the soul, animal spirits, vital spirits, and cosmic consciousness.

Vesalius created the first detailed atlas of human anatomy; it was called *De humani corporis fabrica libri septem*, or *Seven Books on the Structure of the Human Body*. It was an amazing tome and was the first of its kind, even including drawings of dissected human brains.

Vesalius delicately suggested that perhaps the brain is where the mind lives, and perhaps the soul is part of the mind. This was a radical theory for the time.

Afraid of causing an uproar, Vesalius was careful not to push his ideas on anyone. Privately, though, he concluded that the soul doesn't emanate from the heart, trickle into the skull, and derive its facilities from the brain.

Demons, Witches, and Astrology

Only a few hundred years ago, in 1600, most people still believed that the soul was immortal, invisible, and nowhere and everywhere at the same time. (In fact, many people still believe these ideas today; they remain common notions of many religions.)

But in 1600, most people also thought that the heart controls the brain, the soul, the mind, and the entire body. They still believed that each of us is linked to the four elements of earth, water, fire, and air. And they believed in the cosmic consciousness, a vast collective entity made up of souls, stars, the moon, and the sun. They also figured that demons could make men go mad.

Keep in mind that these ideas were widespread only a few hundred years ago. It is easy to see how people who believe in cosmic consciousness and breathing in souls that live forever might also believe in demons, as well as witches, ghosts, and ghouls—and psychokinesis and mental telepathy. It's all connected in the cosmic consciousness, the vast intelligence far beyond anything we can imagine. Even today, people still believe these ideas because they want to feel connected to one another and to the universe. They feel healed and nurtured by thinking they are part of the vast machinery called the universe. Regardless of what science tells us, people will always want to feel connected to one another, to a higher being, to the entire universe. We want to find aliens on other planets. We do not want to be alone.

In the 1600s, doctors cured the sick by using astrology to figure out when they should flush the bad spirits from their patients. Doctors also relied heavily on laxatives, vomiting, and bleeding, all in an effort to flush the demons from the sick patients. It was a common practice for medical doctors to try and restore the natural balance of the four elements in the body, and to do so, they used horoscopes, charms, amulets, religious sermons, and prayers.

Gaia and Natural Magic

Things got even stranger. People started thinking more and more about religion, philosophy, and the soul. In fact, fairly recently they came up with a new philosophy, in which the entire world is a living thing called Gaia. According to the Gaia theory, first proposed by the research scientist James Lovelock in the 1960s, the human soul is simply part of a vast cosmic soul. The cosmic soul runs every-thing: it makes some people happy, and ruins other people's lives, it makes children sick, it cures the elderly, it raises the dead, it gives people ideas, and it influences both major and minor events. This new philosophy was called natural magic.

As you might guess, the official church really didn't like natural magic, which it felt was tainted with pagan religion and witchery. The church maintained that the soul is immortal, invisible, and unique in each person. The church also maintained that God is an omniscient presence with infinite knowledge. There are people who believe that we do not determine our own fates, that all is deter-mined for us prior to our births by this infinite and vast presence. There are others, of course, who believe that we do indeed control our own destinies, and there are even some who believe in a com-bination of the two.

The Mind-Body Connection

It's quite possible that our physical brains are the same as our minds and our thoughts. The soul may simply be a part of the physical brain. The neurocircuitry in the brain may combine in highly intri-cate ways to form our thoughts; another entity, a spiritual mind, a soul, may not enter into the picture at all. As with most everything, there are people who believe that the brain and the mind are one, and there are others who believe the opposite.

Regardless, it is clear that our thoughts are intimately connected to our brains and bodies. If we touch a hot stove, our minds feel pain. If we jump into a freezing lake, our minds feel cold. If our physical senses detect anything through smell, touch, flavor, hear-ing, and so on, our minds instantly know what we're feeling. Our minds receive steady information from our senses, and we contin-ually modify our thoughts based on these senses.

For example, let's suppose that you're sitting on your back patio, reading this book. The sun is warm, you're half asleep (but not from reading this book), and suddenly, you hear a loud crash from the road in front of your house. It sounds as if a trash can has fallen over. Your senses are jarred. You feel a chill rather than the heat of the sun. You're no longer half asleep. The physical world outside your body has affected your senses (hearing, in this case), thus causing your mind to consider new thoughts.

Obviously, this entire scenario can go in reverse, and your mind can affect your physical environment. Let's say, for example, that you get up from the lounge chair, put down this excellent book, shake your head to clear your senses, and walk around the house to investigate the loud noise. You move quickly out of fear that someone might be hurt and need your help. Your mind has affected your physical body, as well as your environment. Your arms and legs are moving rapidly, your head is turning, your eyes are looking for evidence of the loud noise, and the ground beneath you is being trampled by your feet. The grass is being pressed down, possibly killed; insects are being smashed; birds are flying away. And in the backyard on your patio, dew drops on this fine book and splotches the print.

Is it too far of a stretch to think of our minds affecting physical reality in more dramatic ways, such as the ways demonstrated by Carrie? When does matter act upon our minds, and when do our minds act upon matter?

Let's continue our patio scenario. A loud crash occurs and sends sound waves into your ear drum. Your ear drum starts vibrating, and it sends information into a fluid in your inner ear. This fluid triggers some electrical impulses to zoom down your auditory nerves into your brain. It is at this point that your brain registers that you have heard a loud sound. You are thinking about the loud sound, wondering what caused it, whether people are hurt, whether a truck's about to explode. You're thinking about possible consequences—can you call the fire department quickly enough, can you get the ambulances there on time? Your mind, which was half asleep only a moment earlier, is now grinding at top speed. It starts making decisions, such as forcing you to run around the house to investigate the noise. Emotions start to gurgle in your mind. You're worried that someone is hurt, possibly a good friend and neighbor who is very dear to you. You start to feel sad. Your feet pick up the pace. You must get to the front of the house more quickly, for if your friend is hurt or dies, the guilt will consume you. Your conscience is in play, your soul.

We each would reach this scenario in slightly different ways. Our consciences may differ, our ability to empathize, to react quickly, to consider the possible outcomes of the noise. Our souls might react differently. A lazy and self-indulgent bum, for example, wouldn't care if his friend has been crushed by a fallen helicopter. An insane lunatic would be thrilled to think that a truck has turned over on top of the little old lady across the street; he might toss back a beer while sauntering around front to check out the bloody gore.

To a material scientist, only the hardwiring of the brain, the neurochemistry, determines whether you react as you would, as the lazy and self-indulgent bum would, or as the insane lunatic would. Give the bum psychiatric drugs, and he might conform to what we consider normal. Give the insane lunatic lots of psychiatric drugs, and he might turn into a lazy and self-indulgent bum. Who knows? What the material scientist might maintain is that everything we do boils down to neurotransmissions. There is no mind, no self, no thought process that can't be explained by physics.

Free Will versus Fate

We prefer to think that we have some control over ourselves. If all we are is a pile of circuits, then what's the point of even trying to become better people? Our fates don't belong to us. Most of us therefore believe that we have free will, which enables us to decide whether and how we respond to a loud crash. Our souls—our minds—require free will to operate.

Certainly, Carrie uses free will to decide whether and how to respond to the cruelty inflicted upon her by her classmates at the prom. She consciously decides to destroy the place and kill everyone. Sure, their fates are in her hands, but if you think long enough about it, had they done things differently long before the prom, their fates would have been quite different and under their control.

If everything in the universe is linked by cosmic consciousness and Carrie is able to set fires and kill people simply by wishing it so, then her free will enables her to unleash this cosmic power when she needs it. If matter can act upon her mind, then her mind can act upon matter.

How does free will function within the framework of physics, which has rigid deterministic laws? If neurotransmissions are responsible for everything we think and do, then how do we influence our own thoughts and actions using free will? At what point do we take over, with the neurocircuitry responding to our every wish and desire? And if free will is not part of the physical circuitry, then just what part is the *we*?

Circuits supposedly operate in fixed ways: if *x* happens, do *y*; if *y* happens, trigger *z* and stop *a* from working. Is it possible that our minds plumb the depths of brain cells and nerves to create the neurotransmissions that enable us to create new ideas, to make decisions based on personal morals and private memories? Is the mental truly part of the physical brain, or does the mental operate in conjunction with—but separate from—the physical brain?[5]

If free will is our ability to be individual selves with individual souls, we might assume that free will is an aspect of what is thought of as consciousness. It is generally assumed that animals do not possess this consciousness. Basically, nobody knows whether monkeys are conscious and whether we're torturing them in laboratories, and whether worms, artificially intelligent computers, and dogs are conscious. An artificial intelligence, for example, might possess consciousness and the ability to know right from wrong, to realize that it is an entity, an individual, a self. The fully sentient computer might realize, "I think, therefore I am." The debate about abortion may center on issues of consciousness: just when is a baby considered conscious, an individual self with an individual soul? Possibly, a human baby is conscious six months before birth, or maybe two months before birth, or maybe only minutes before birth. Maybe the human baby is conscious only during and after birth. Nobody knows for sure. And who's to say that our dogs and cats aren't conscious and operating with free will? It's possible that consciousness formed a long time ago in the evolving brain.

The Dualistic Theory

Remember that in early times and as recently as the 1600s, people believed that the soul is a life force that exists in the lungs, the heart, the liver, or the blood. This life force is literally what gives us breath and lets us move and think. In the New Testament, the soul is very much the same thing as the self and the mind, with the *Catholic Encyclopedia* stating that the soul is "the source of thought activity."[7] In short, the soul is an entity. This entire notion is the dualist theory of the body and the mind (soul) initially developed by Descartes.

The dualist theory tells us that we actually consist of two parts, a soul or mind, and a physical body. The soul/mind lives inside the body, which serves as a host receptacle. Descartes pondered that the soul might reside in the brain's pineal gland, where the ephemeral mind would then interact with the physical body.

The dualist theory leads spiritual people to believe that they can release their souls from their physical bodies, hence releasing their souls from physical constraints or prisons. This is a major reason that shamans and other spiritual seekers study, starve, and pray—to release their souls from the prisons of their bodies. Death is the ultimate release of the soul, whereby the soul returns to cosmic consciousness and ultimate freedom.

If all of the previous is true, then we could conclude that the soul can reside outside the body. Now if the soul can reside outside the body, then who's to say it can't hang out near someone else and give that person's soul a few thoughts, as in mental telepathy?

While the soul can reside outside the brain (but, while we're living, hopefully not outside the body), it is connected to the body through the brain. Some modern philosophers refer to our bodies as machines or engines, and they think of the invisible and weightless soul as the force behind the engines. If the human body is a machine, the soul is "the ghost in the machine."[8]

If the soul is an invisible ghost, we're left to wonder where it is. How does the air around us contain the billions of souls that were once attached to live humans? Where are our souls before we're born? Does the soul exist, or does it come into being the moment a human is born?

Early Roots of Neuroscience

After Descartes, a doctor named Thomas Willis started working on theories about the brain, the nerves, and the soul, among other things. Willis was one of the founders of neuroscience, and, like Vesalius, he dissected the brain repeatedly to learn more about what was inside the skull. Willis figured that blood enters the brain, where the natural spirits are distilled from it. Remember, earlier doctors also believed that vital spirits turned into animal spirits at the base of the brain, and then these animal spirits flowed into the mind. Willis added to this theory. He said that once inside the brain, the animal spirits traveled through the neurocircuits and escaped from the body as vapor.

Willis was the fellow who discovered that the human brain consists of regions. At the base of the brain is marrow, where the spinal cord connects to the brain. We now call this part of the brain the medulla oblongata. Willis noted a ball shape above the medulla oblongata, and this ball structure is the cerebellum. Above both the medulla oblongata and the cerebellum is the cerebrum, which has two hemispheres. Willis continued exploring the parts of the brain by completing more detailed dissections.

By 1663, Willis and his team completed their research. Willis coined the term *neurologie*, which meant "doctrine of the nerves." He devised additional theories about animal spirits and how they move through brains and nerves, and he and his team created new brain anatomy diagrams.

The new diagrams were included in a book called *The Anatomy of the Brain and Nerves*, and Willis had to ask the church for permission to publish his research. His book was far more than a map of the brain or a series of drawings. It had far-reaching consequences. It affected religion, philosophy, and everything mankind thought about itself. In addition, the book basically started what later would be called the science of neurology.

Despite Willis's tremendous achievements in science, including the founding of neurology (no small feat), he still believed that the soul consists of animal spirits that flow around the brain and the nervous system. He further believed that nonhuman animals have souls as well, and that their souls are based on the sizes and the structures of their different brains. The animal spirits of a tiger make it move and act differently from the animal spirits of a monkey. Willis further adhered to the notion that only human souls have sensitivity, rationality, and consciousness. He did not think that the soul consists of matter. In fact, he felt that while people sleep, their souls must rest and recharge by withdrawing from their brains.

As for when babies get souls—before, during, or after birth—Willis said that God places a soul into the baby's brain before birth. As the church taught, the soul is immortal and exists after a person dies.

Then Willis expanded on the more philosophical ideas surrounding the soul. He decided that there is a rational soul and a sensitive soul. It's possible that he maintained certain church-related beliefs in parallel to his scientific research because he felt compelled to protect himself, his team, and his work from public outcries. Or maybe he really believed what he claimed about the soul; one can only guess.

At any rate, Willis asserted that the invisible, immortal rational soul lives in the part of the brain called the corpus callosum. This is the soul that lives after a person dies.

As for the other soul, the sensitive soul, Willis said that it consists of matter. It is not invisible. It controls details about everything that is unrelated to our senses.

The sensitive soul obtains impressions and images, and it relays this information to the rational soul. It is the rational soul that gives a person free will, reason, wisdom, compassion, and ideas. The rational soul rules the body. The sensitive soul operates the body and makes sure that all the animal spirits are flowing around correctly.

Willis further believed that the two souls bicker for power inside the brain. Every now and then, they bicker so much that the brain gets sick.

Let's say your sensitive soul gets tired of being bossed around by your rational soul. The sensitive soul gets depressed. You get depressed. (Imagine a modern psychotherapist trying to deal with depression under these circumstances.)

The rational soul continues to browbeat your sensitive soul, which becomes increasingly depressed to the point of wanting death. The acute illness of the sensitive soul starts to affect your rational soul, which becomes sick, too. If your rational soul gets sick, it's over, pal: you go insane, with your rational soul producing hallucinations and delusions. Willis penned a second book of neuroscience, this one devoted to what might be thought of as early psychiatry. His second book, which described his theories about the soul, was called *Two Discourses Concerning the Soul of Brutes*.

Modern Neuroscience

Neuroscientists today no longer refer to vital spirits and animal spirits. Instead, they call electrical impulses that course through our brains *chemically based neurotransmissions*. These neurotransmissions pass signals among the neurons, or brain cells.

While neuroscientists are still trying to map what happens from neuron to neuron as we see objects and think about what to do when we hear loud explosions, little is known about the human brain. There's a lot more to learn, and we've barely figured out what makes us tick.

Even today, the idea of consciousness or self remains elusive. The Consciousness Research Institute (visit www.deanradin.com/CRL.htm for more information), in California performs research into psychic powers and the role of consciousness in the material world. Studies focus on psychic powers similar to those experienced by Carrie—mind-matter interactions—as well as on clairvoyance, precognition, and distant healing.

A common parapsychologist argument is the following: we don't know that psychokinesis isn't science until we can prove that it isn't science. This is an inverted sort of argument, that just because we can't prove *x* doesn't mean that this particular *x* isn't a scientific truth. Just because we can't prove that 5,239 gods exist doesn't mean it isn't scientifically accurate to state that they do indeed exist. Just because we can't prove that elephants think in English doesn't mean that it isn't scientifically accurate to state that they do think in English. And so forth.

These arguments about psychokinesis are not particularly convincing. Nonetheless, many of us think that there are many things we don't know about the human mind and the world around us, and one of those things might be that the mind is capable of psychic powers. Someday, a Carrie may come along whose thoughts can collapse buildings and start fires.

Carrie and Charlie: A Combustible Duo

Similar to *Carrie*, *Firestarter* (1980) features a young girl, Charlie McGee, who has pyrokinetic abilities, which means that her thoughts can start fires. (Like *Carrie*, *Firestarter* was also made into a movie, with Charlie McGee portrayed by Drew Barrymore, in 1984.) In each case—*Carrie* and *Firestarter*—the control of the psychic powers is based on hormonal cycles. For Carrie, the hormones in question are directly related to the onset of her menstrual cycle. For Charlie, the powers are associated with the development of her pituitary gland. For each girl, home life is abnormal: Carrie's mother is a religious maniac, and Charlie's parents are no longer with her. For Charlie, an intrusive government eventually kills her mother, leaving her with only her father to watch out for her. Because Carrie's mother is insane and the adults at her school don't intervene to help save her from abusive and violent bullies, she is eventually forced to use her supernatural powers to save herself in the only way she knows how.

Before Charlie McGee was born, the government used her parents in secret drug experiments called Lot Six. The government gave her mother and father powerful hallucinogens, and by the time they conceived Charlie, their genetic makeups had been altered by the drugs. Charlie's pyrokinetic powers were the result of genetic mutation.

A covert government agency called the Shop, which we assume is like the CIA, wants to use Charlie and her powers, and hence they will pursue her no matter what the cost, even killing her parents to get to her. The government thinks that when Charlie becomes a teenager she will be able to destroy the planet simply by wishing it so. She will be able to destroy whomever and whatever the Shop wants her to destroy. Her pyrokinetic abilities are worth anything and everything to the covert Shop.

After being captured, Charlie undergoes pyrokinetic experiments at the Shop's compound in Virginia. Her father is with her, as is a pseudo–Native American named John Rainbird, a psychopath who wants Charlie's powers for his own use. Rainbird does everything he can to pry Charlie from the government, as he hopes to obtain her pyrokinetic powers, elevating himself to the afterlife of a god. He also wants to watch her die. Basically, he is yet another psychotic adult in Charlie's life. In the end, to save her own life, Charlie burns down the compound and escapes.

As with Carrie's telekinetic powers, Charlie's pyrokinetic powers are vaguely possible in real life, but they remain unproven. Parapsychologists explain pyrokinetic powers as the ability to excite an object's atoms, generating sufficient energy within the object to fuel a flame. This is generally the stuff of comic books, such as the Human Torch (Marvel Comics), who can spontaneously ignite himself in flames that enable him to fly.

The psychic ability of pyrokinetics is related to the notion of spontaneous combustion, which usually refers to the self-ignition and burning of a person or an object. Suddenly the person or object bursts into flames for no apparent reason. This is possible to a degree because some substances actually do ignite when they hit air. Caesium, for example, is a soft silver-gold metal that is liquid at or near room temperature. When combined with water, it forms a solution of caesium hydroxide and hydrogen gas. This reaction is extremely exothermic and happens so quickly that if the caesium and the water are combined in a glass, the glass will explode. Not only does caesium cause fires when it is in contact with the air or water, its hydroxide dissolves flesh and bone.

Silane, a chemical compound of silicon and hydrogen, is a pyrophoric gas at room temperature, meaning that it spontaneously combusts in air at room temperature. Rubidium is also highly reactive, igniting spontaneously, its fire red violet, when in contact with air. So if a person like Charlie can release a substance such as these into the air at a distance, then she can cause explosions and fires. Without an electromechanical device or a wireless transmission to a device located at the place of the fire, it's hard to determine scientifically how the brain can trigger explosions from a distance.

Johnny Smith: The Man Who Knew Too Much

Along with Carrie's telekinesis and Charlie's pyrokinetics is the ability that Johnny Smith of *The Dead Zone* has—to see the future. This novel differs from *Carrie* and *Firestarter*, in that Johnny is not a young girl on the brink of womanhood. Rather, he is an adult English teacher, who, after teaching a class about Edgar Allan Poe's "The Raven," is in a car accident that changes his life forever. Rather than being caused by hormones and glands, Johnny Smith's psychic awakening is due to a five-year coma.

When Johnny awakens from the coma, he has the power to foretell the future. Later, while teaching a young boy about "The Raven," Johnny is able to predict how the boy's life will unfold.

Johnny is normal in almost every way; even his name is that of the ordinary man. Before the accident, he is happy, in love with another teacher, and planning to get married. He is so contented with life, just as it is, that he even turns down an offer of sex made by his fiancée, Sarah, saying that he wants to wait for their wedding night. Then he drives into the night to go home. En route, his small car smashes into a milk tanker. (We won't comment on the obvious conclusion here that had Johnny had sex with Sarah that night, he never would have been on that road, and his small car never would have crashed into the tanker.)

When Johnny finally comes out of the coma, he is no longer ordinary in any way. His job as a teacher is over. His beloved Sarah has married someone else. But most important, Johnny Smith has a new ability: if he has physical contact with another person, he can instantly see that person's past and future intertwined.

As time passes, Johnny's life becomes a nightmare. After all, thousands upon thousands of people need his help. How can he not help a child? How can he not help Sarah, whom he never stops loving? How can he ignore anyone in need?

In the end, Johnny sacrifices his own life to save the world from an evil politician. He knows what the politician will do to people everywhere if he is allowed to gain power.

Physics and Psi Powers

Is it possible to foretell the future, as Johnny Smith does? Here, we're returning to questions about parapsychology and physics as we know them today. For parapsychology and psychic phenomena to be true, we must prove conclusively that our current notions about the world are incomplete.

Some physicists claim that only a conscious entity, such as a human being, can make measurements; and because the existence of matter depends on measurement, then it follows that the existence of the universe depends on consciousness.[9] This harks back to what we wrote earlier in this chapter about cosmic consciousness and all the arguments for and against the idea.

On the flip side, most people know that measurements occur without a conscious being taking the measurement. For example, a camera can measure distance and lighting, and it is not a conscious being.

In today's world, faith in spiritual aspects of life takes a back seat for most people to scientific materialism. Some people, however, including physicists, think otherwise.

For example, Dr. Amit Goswami, a professor of physics at the Institute of Theoretical Sciences at the University of Oregon, believes that our cosmic, collective consciousness is the backbone of reality. Indeed, he believes that cosmic consciousness is more important to reality than matter itself. He decided to ask other researchers, scientists, and authorities about cosmic consciousness.

Supposedly, the physicist Murray Gell-Mann told Dr. Goswami that physicists are brainwashed into thinking that all the fundamental aspects of quantum physics were discovered sixty years ago.[10] It was his feeling, apparently, that we have far more to discover about quantum physics than we think. Many physicists might agree with Gell-Mann, and his statement isn't proof—at all—that he believes in cosmic consciousness.

According to Goswami, the neurophysiologist Roger Sperry, the physical chemist Ilya Prigogine, and the physicist Carl Sagan all believe that everything is made of matter, that there is nothing else to consider and consciousness is a phenomenon of our brains. Finally, the philosopher Karl Popper holds the view, according to Dr. Goswami, that consciousness must be separate from the brain in order to affect it.[11]

From what Goswami reports (and he believes in cosmic consciousness), most scientists believe in physics and scientific materialism as we know them today. To give validity to quantum objects, we must be able to observe them. Because quantum objects have wave properties, they may be in more than one place at once and they may influence other objects that are far away. Without our observations of them, quantum objects would have no value. It is our consciousness that enables us to observe them.

The World as a Machine

In the seventeenth century, when René Descartes visited the palace in Versailles, he was fascinated by the "automata" in the palace gardens. As he watched the automata control the water coursing around the garden, as well as control the music, he thought that the world might be like a huge automata. Perhaps the world is like a machine. As noted earlier in this chapter, Descartes devised the famous theory of dualism, which divided the world into a domain of science and matter (materialism) and a domain of mind and religion.

While medical doctors were establishing the fields of neuroscience and surgery in the mid-1600s, Isaac Newton proposed theories that established Descartes' "world as machine" ideas as scientific truth. Newton created the principle of causal determinism. Simply stated, it means that given the laws of motion and the facts about where objects are before they move and how fast they're moving, we can predict the exact locations of the objects.

Space-time and Material Monism

As we entered the nineteenth century, classical physics had two main principles: strong objectivity, or the separation of science from the mind, and causal determinism. Albert Einstein added the theory of relativity, which, among other things, suggested that the highest velocity in the universe is the speed of light, which is 300,000 kilometers (or 186,000 miles) per second. Einstein's theory further suggested that all interactions between objects in space-time must be local: objects travel one bit at a time with a finite velocity. This idea is often called locality.[12]

As scientists explored the separation of science and matter from the mind, they came up with the notion of material monism. This means that everything in the universe, including our minds and consciousness, is composed of matter. Because nobody knows how to prove the connections between mind, consciousness, and matter, scientists use the term *epiphenomenalism* to describe the derivation of consciousness from matter. In general terms, epiphenomenalism hypothesizes that consciousness is a series of properties of the brain, which is composed of matter.

All of this fascination with consciousness, the soul, the mind, and matter leads to notions about mental telepathy, pyrokinesis, healing the sick, raising the dead, and other paranormal abilities. Yet another offshoot of these scientific, philosophical, and religious suppositions is the idea that we can foretell the future, just as Johnny Smith is able to see what will happen before it occurs.

Ancient and Modern Oracles

The ability to see into the future and warn people about terrible events is something that mankind has believed in for thousands of years. People of all cultures believe in this psychic ability.

Consider this childhood toy: the Ouija board. The board displays all the letters of the alphabets along with words such as *yes*, *no*, and *maybe*, on it. To receive messages from the Ouija board, a person puts his or her fingers on a three-legged device called a planchette. The person's friends may also place their fingers on the planchette, and then someone asks a question of the board. Everyone waits for some outside force to move the planchette, and then the planchette moves by itself and provides an answer, either by spelling it out or pointing to the words *yes, no,* or *maybe.*

In ancient Greece, people often asked oracles for answers to their problems. If you wanted to learn whether you were finally going to have a child, whether your brother survived a war, or whether your fiancé had a hideous disease, you could ask the famous oracle at Delphi. After giving offerings to the gods and begging priests and priestesses to help, you'd go away and wait. Eventually, the priests and the priestesses would consult the oracle and supply you with its answer.

Today's oracles are called fortunetellers, tarot card readers, and, more commonly, con artists. Nancy Reagan, the wife of President Ronald Reagan, had a private fortuneteller who regularly gave her advice, even about the country and its policies. The daily newspapers are full of horoscopes and astrology fortunes. At Chinese restaurants, you receive fortune cookies after your meal.

A potent and intricate method of seeing into the future is the Chinese method called the *I Ching.* It's been around for thousands of years and is also called *The Book of Changes* or the *Chou I.* The Confucians studied the *I Ching* during the last period of the Chou era, and at that time, with most books banned, the *I Ching* was one of the few books that the government allowed people to study. In fact, in 140 B.C., the imperial academy excluded all non-Confucian texts, and the *I Ching* became doctrine, with its own chair of study.

Over time, the *I Ching* became a volume of "sacred scriptures inspired by divine revelation. The reason seems to lie in the concentration of divine as well as temporal power in the person of the emperor, in China as well as in other oriental societies. The emperor was not only the sole source of political decisions, he was also the Son of Heaven, the representative of the deity among men."[13]

Similar to cosmic consciousness, the *I Ching* views the universe as a natural whole in which change is continual yet connected. This enables the *I Ching* to give advice and tell you what to do if you hear a loud noise, want to control your destiny, or want to see into the future. The *I Ching* is not a simple book to study; it takes many years of concentration to use it correctly in hopes of seeing into the future and guiding human thoughts and actions so that they are aligned with the cosmos.

The *I Ching* consists of symbols in the form of sixty-four hexagrams. Each hexagram is composed of six horizontal lines, some of which are solid, while others are dashed. Outer hexagrams form a circle around inner hexagrams that form a square. Each hexagram is made of a pair of three symbols, called trigrams. Each trigram has a special meaning, and to learn the meanings, a person studies the *I Ching*, which interprets all the meanings in a series of commentaries. Everything is intertwined, and careful interpretation is required of the commentaries of all trigrams in connection to one another.

Johnny Smith has a much easier time seeing into the future. He just suffers a lot, seeing into people's futures and minds, watching them feel pain and endure agony. If he had to consult a ouija board or the *I Ching*, he would turn away from foretelling the future in an instant. As it is, he considers his psychic ability to be a curse.

Ted Brautigan: The Reluctant Good Samaritan

Like Johnny Smith in *The Dead Zone*, Ted Brautigan in *Hearts in Atlantis* (1999) can also see the future. Brautigan avoids touching people because he doesn't want to see into their minds and see their futures. Yet he helps children to the point where Bobby Garfield's mother, Liz, thinks he may be a pedophile. For example, when Ted tries to fix young Carol Gerber's dislocated shoulder, Liz thinks he is touching the girl far too much.

When Ted touches someone, he can sense what's happening to that person. He opens a window into their minds, but beyond that, he knows what will happen based on what they're thinking and what they will do. He possesses the gifts of mental telepathy and prophecy. In addition, when he opens the window into someone else's mind, he passes his psychic powers to that person—if only for a moment.

Cell: The Ultimate in Groupthink

In King's book *Cell* (2006), mental telepathy and cosmic consciousness are once again the supernatural science themes. Here, cell phone towers are transmitting signals that wipe brains down to their primal instincts. It's never explained how the cell phone towers erase people's brains—possibly, terrorists have infiltrated cell phone transmissions. At any rate, along with the brain wiping, the cell phone signals give people the ability to transmit brain signals to one another. They have mental telepathy, in other words, which later binds them together into a *Star Trek* Borglike hive that King refers to as the Flock. There are huge Flocks all over the country in all major cities. These Flocks sleep in malls, stadiums, and other areas that are big enough to accommodate them. Members of the Flock sleep side by side, with music blasting from their open mouths, and as they sleep in this way, they communicate with one another using mental telepathy. Periodically, each person reboots his or her wiped brain using cell phone signals. When not rebooting their brains with cell phone signals and using mental telepathy to communicate with one another, members of the Flock savagely murder normal people. Remember, the most base instincts of the brain are the only parts of the brain left.

Members of the Flock are connected as if they are one organism. Because so many people these days use cell phones, nearly everyone becomes a Flock member, and eventually, the Phonies outnumber the people who weren't using their phones when the first signal was transmitted, or Normies. The Phonies are able to get into the Normies' dreams and brains using mental telepathy, and they get the Normies to speak words on the Phonies' behalf and to do things that favor the Phonies' lifestyle, if we can call this a lifestyle, over that of the Normies.

A computer worm infiltrates the cell phone signals—again, we're not sure how this happens or who instigates the worm. When the Phonies reprogram their brains at night, the worm keeps mutating, causing some of the Flock to want to be Normies again. In the end, it's the child Jordan who theorizes that if a person holds a cell phone to his ear and dials 911, the new signal will wipe the brain again. Just as a computer hard drive saves old programming as a survival instinct, or so Jordan explains, the wiped brain might have saved the old "programming" as a survival instinct. And because the worm wipes the brain in conjunction with a 911 call, the brain's original programming may kick in. The original minds of the Phonies may be restored.

We've already discussed psychokinetic powers, which include mental telepathy. We've suggested that a cosmic consciousness or a universal "oneness" may be responsible for these psychic powers. Some supporters of mental telepathy put forth quantum theory as a possible explanation of psychic power. We've previously touched on this idea, but let's return briefly with a focus on telepathy. If the human mind consists of quantum and electrical impulses, and if the mind can pick up on quantum fluctuations that are generated by other minds, then it's somewhat conceivable that minds can communicate directly.

On the other side of the debate are scientists, who point out that quantum mechanics deal with subnanometer entities, yet the brain deals with much larger entities and impulses.

In response to this opinion, people who believe in mental telepathy use the argument presented earlier, that it's possible because it hasn't been disproven and there may be entire areas of physics we now nothing about. As noted, this is a somewhat inverse argument and doesn't hold up as scientific proof of anything. Sure, anything's conceivable, and pigs may fly, but until repeatable proof exists for something like mental telepathy, it's hard to call it a science. We can call it a pseudoscience, and we can say that it's vaguely possible—for pretty much anything is possible.

John Coffey: An Unlikely Healer

In *The Green Mile* (1996), a poor black man, John Coffey, is convicted of raping and killing two white girls. He stands well over seven feet tall and weighs 350 pounds. While he may be huge and scary to those who condemn him, he is an innocent man, wrongly accused and convicted.

While on death row, John displays the amazing ability to heal the sick, ease suffering, and punish evil that lurks in the hearts of bad men. For example, he cures the prison warden's wife of brain cancer. He also restores life to the dead pet mouse Mr. Jingles. John pulls the suffering from other people into himself, then expels their illness and misery from his own body in the form of a visible demonlike thing.

One of the prison guards, Paul Edgecomb, realizes that John is truly a nice man who is innocent of all crimes. Paul is helpless in the face of the system that has condemned John and actually must serve watch during John's execution.

Just as Johnny Smith in *The Dead Zone* realizes his powers when he comes into physical contact with other people, John Coffey in *The Green Mile* realizes his powers when he touches someone else. In the latter case, Coffey connects to human sin and evil when he touches another person. Like Smith, who is exhausted from requests for help and experiences of human misery, Coffey becomes fatigued from the human evil to which he is exposed day in and day out. Both men possess psychic gifts that let them help others, yet both men find their gifts to be excessive burdens. And like Smith, Coffey can transfer his visions to someone else during physical contact. He enables Paul to see the identity of the two girls' real murderer, and he lets Paul see the human sin and misery that he, Coffey, is forced to experience all the time.

In reality, cultures around the world have believed in faith healers and medicine men for thousands of years. The notion that one man can cure the sick and return life to the dead has been around for a very long time—since prehistoric times.

A shaman, for example, is a medicine man with magic-religious powers who cures human suffering by forming relationships with spiritual entities. The shaman goes into a trance, or spiritual state, and asks the spirits how to heal the sick, raise the dead, and save the tribe or the nation.

Medicine men and shamans have existed all over the world throughout history. The word *shamanism* comes from the Russian evolution of the Tungusic word *saman*. We could list many examples of medicine men and shamans in all cultures. One such example might be the Tatar people, who used the shaman for almost everything. For example, to cure a sick child, the shaman would hold a séance to try and bring back the soul of the child. The séance could last six hours, maybe more, during which the shaman would go into a trance, traveling to the lands of the spirits. The shaman would search for the sick child's undamaged soul and ask the spirits for a way to heal the child's illness.

In keeping with the universal ideas throughout history of cosmic consciousness and the interconnection of human souls, the shaman travels from one cosmic region to another for advice and help. He is able to communicate on a cosmic plane via the cosmic consciousness.

We can think of John Coffey as a modern-day medicine man or shaman. He heals the sick, and he can bring creatures back to life. Like the Tatar shaman, who cured sick children and brought creatures back to life, it's possible that John Coffey can do the same thing.

Yet there are no experiments or verifiable records of a shaman bringing people back to life. Whether this is truly possible is unknown, just as it's unknown whether a man like Johnny Smith can see into the future and possesses mental telepathy, whether a girl like Carrie can kill and destroy simply by thinking about the acts, whether a girl like Charlie can set fires using her mind, or whether people can connect to the cosmic consciousness as they do in *Cell*. This is one reason that reading Stephen King is so fascinating. In all of his books, he asks the universal question, What if?

ON THE HIGHWAY WITH STEPHEN KING

"Trucks"

Someone must pump fuel.
Someone will not be harmed.

—"Trucks"

In Stephen King's world, trucks can come to life, gather forces, and kill people. In our real world, vehicles are becoming "smarter" in that they have global positioning software navigation and other computer controls. But does this imply that someday cars and trucks will actually turn us into gas-pumping slaves?

"Trucks"

Most of King's early short stories were collected in the book *Night Shift* (1978), his first short story collection and fourth book published by Doubleday. One of those short stories was "Trucks."

Though it is considered one of King's less memorable works, "Trucks" was twice made into a movie. The first version of the story, called *Maximum Overdrive*, not only featured a screenplay by King inspired by "Trucks," but was also directed by him. The film debuted in 1986 to dismal reviews and less-than-spectacular box office success. It was the only film King ever directed. In 1997, a made-for-TV movie version of "Trucks" with a teleplay by Brian Taggert debuted to similar bad reviews. The uninspired story line somehow managed to connect Area 51 with the events. The TV movie was later released in theaters overseas. Both versions of "Trucks" are available on DVD. During the last few years, *Maximum Overdrive* has become somewhat of a cult film, gaining a loyal following among film fans who love its "drive-in" movie–style atmosphere and dialogue.

The basic concept of "Trucks" is simple. A small group of people is stranded at a roadside diner. Outside is a gathering of trucks that somehow have become intelligent. The trucks are killing all humans in the area. No one has any idea what caused this event, nor are any explanations offered. One character wonders about nuclear power or electrical storms. There's no mention of a comet passing through the atmosphere or Area 51, the reasons offered in the two movies.

When the population of the diner is knocked down to five survivors, the trucks finally make an effort to communicate. A horn honks.

"That's Morse!" a kid, Jerry, suddenly exclaims.

A trucker looks at him. "How would you know?"

The kid goes a little red. "I learned it in the Boy Scouts."[1]

No one wonders how the trucks know Morse code, but logic is not a major force in the story. The trucks are running low on gas. If the humans pump them full of gas, the trucks will let the humans live. After two more victims (Jerry and the trucker) are killed, the three survivors—the narrator, Jerry's girlfriend, and the counterman at the diner—begin the tedious job of pumping gas into the trucks. The narrator, taking a break, wonders whether this might be the end of the human race. He cheers up when it occurs to him that trucks can't reproduce—only to realize that automated assembly lines have changed even that. Humanity, he concludes, is doomed.

On that cheerful note, the story ends. As mentioned, there's not enough logic to even start an argument about the story being believable. The short story version endows only trucks (and a single Greyhound bus) with minds and intelligence. The two film versions add such disparate monsters as living vending machines and electric knives. In the print version, only trucks come alive and maybe, it is hinted at the story's conclusion, airplanes. No cars, no vending machines, no lawn mowers. It's as if somehow trucks, because they are bigger than most other machines, have more intelligence or cruelty than Cameros and Cadillacs.

When desperation sinks in, and the machines finally realize they need humans to pump gas or they will run out of fuel and die—or at least not be able to operate—they communicate by Morse code. Recently, newspapers have run articles mentioning how Morse code has nearly been forgotten.[2] It's pretty amazing that one teenager remembers the signals in the story and even more incredible that he can recall the letters after watching a half-dozen people murdered right outside his shelter. Still, we accept author's license, especially in a tight spot in a story. But we can't help thinking that trucks without any sort of backup computer memories might not know Morse code right after they suddenly come to life. Not even a computer virus works that well.

There's something about the basic story line of "Trucks" that obviously resonates with horror fans. One of the highest-rated TV movies of all time, *Duel*, features Dennis Weaver as a driver being pursued by a relentless semi-trailer truck in which the driver is never seen. Movies with similar themes have also done well in the past. It's clear that a sizable number of people find the battle between man and machine exciting, especially when the man isn't a macho character like the Terminator but an ordinary short-order chef at a diner on a highway leading nowhere. It's a twenty-first-century David versus Goliath story, with the outcome no longer so clear.

"Trucks" resonates on another, deeper level, one that isn't immediately apparent but is in the story nonetheless. In 1917, the noted horror author Arthur Machen wrote a short novel called *The Terror* (1917), in which a series of unexplained murders takes place across the English countryside during World War I. No explanation is given for the murders, none of them are solved, and they all involve animals. The point is subtle but clear by the end of the novel. The creatures of the earth are angry with man for devastating the world with war and are striking back.

A similar theme was the basis for J. T. McIntosh's 1950s novel *The Fittest* (1955). That book describes a future where domestic animals turn feral, the result of a science experiment gone bad. The novel is set at a small farm outpost where humans are forced to battle wolves, rats, foxes, and other commonplace animals that have suddenly turned into vicious killers.

Take that concept one step further and the result is "Trucks." Animals aren't as commonplace as they were a hundred years ago, even fifty years ago, while trucks are everywhere. They number in the tens of millions. If somehow the Earth were to rise up in protest of man's treatment of the planet, as "Trucks" so amply demonstrates, trucks would serve as the perfect tool of divine vengeance.

Is this too outlandish a meaning for a simple short story? Not when you consider this short exchange between the narrator and the trucker:

> "What would do it?" The trucker was worrying. "Electrical storms in the atmosphere? Nuclear testing? What?"
> "Maybe they're mad," I said.[3]

It's extremely doubtful that our trucks will be coming to life and attacking us any time soon. Yet cars are getting smarter and smarter all the time. We can now talk to our cars and tell them what to do, and they will follow our instructions and even report when they are finished. Cars can send messages to repair shops regarding when they need replacement parts. Soon, cars will exist that can be programmed to drive on their own to places all over the map without any person in the driver's seat. It's all coming tomorrow, courtesy of artificial intelligence, most often called AI. Which leads us directly back to the basic concept of "Trucks." Will cars and trucks come alive sometime in the near future? Is mankind foolishly manufacturing our own successors?

It's an idea that's been with us since the beginning of the Industrial Revolution. Long before the invention of the computer, people were worried about the notion of thinking machines turning on humanity. One of mankind's deepest fears is that the computer revolution might be more about revolution than it is about computers. Does the ongoing research into artificial intelligence spell doom for the human race?

Early Concepts of Artificial Intelligence

Stated in the simplest terms, artificial intelligence is intelligence demonstrated by a nonliving being. In most cases, AI is associated with computers, robots, or androids. A computer is defined as a machine that manipulates data according to specific instructions known as a program. Originally, computers could be programmed only to do one task at a time. In today's world, computers can do vast numbers of tasks all at once; however, computers at present do not possess artificial intelligence.

Frankenstein

The earliest novel dealing with an artificially intelligent being is Mary Shelley's famous book *Frankenstein*, published in 1818. The story is much different from most of the movie versions familiar to today's audience. It tells how the brilliant researcher Victor Frankenstein creates a man out of parts of dead bodies and brings it to life by alchemy. The creature is horrible looking. Frankenstein loathes his creation and abandons it. The monster is innocent of the world's ways but is not evil. Only after being rejected again and again by the people it meets, as well as by its creator because of its horrifying appearance, does the creature finally turn to murder. It does so logically, killing the people who mean the most to Frankenstein. At the book's end, Victor dies and the monster, overwhelmed by grief, vows to commit suicide. In this case, the artificial intelligence of the monster is actually greater than that of the people it encountered.

The Frankenstein monster was the earliest example of an android, a creature considered by early science fiction writers to be an artificial organic being, a "synthetic" man. A number of early horror and fantasy stories used androids as the villains. Because they were made artificially instead of being the product of the union between a man and a woman, such creatures were thought to be created without souls. Being without souls, these evil entities were considered incapable of human emotions. The term *android* has been borrowed by many horror writers, including Stephen King, ever since.

R.U.R.—*Revolt of the Robots*

Robots are machines, often in the form of humans, that are programmed to perform physical actions. The word *robot* came from the play *R.U.R.* (Rossum's Universal Robots), by the Czech playwright Karel Capek, produced in 1921. In the play, robot beings in human form are used as cheap labor until they rise up in revolt and kill their human masters. *R.U.R.* helped to popularize the notion that robots should be designed to look like humans, which made little sense in real-life situations.

Robots of all sizes and forms are used in modern industry. Machines are built specifically to do certain jobs in and are designed in a manner to perform those jobs in the best way possible. Though *R.U.R.* is told in a very different style from "Trucks," their themes are very similar.

The Industrial Revolution, with its flurry of machinery, popularized the idea of mechanical men, and these robotic creations appeared in everything from Frank Reade dime-store novels to the Oz books by L. Frank Baum. In these early fiction works, robots were usually portrayed as friendly to mankind. It wasn't until Capek's play that robots were thought of as being self-centered and malevolent. This notion that artificial men were evil grew with the

1931 release of Universal Picture's movie version of *Frankenstein*, starring Boris Karloff. In the film, the artificial man is originally shown as being ignorant but not evil. Audiences felt little sympathy for the creature, however, and he was considered by most to be a monster.

More Uppity Androids and Robots

In his novel *The Moon Pool*, published in 1919 by Putnam, the popular author Abraham Merritt introduces an artificial intelligence known as the Dweller, with near godlike powers, which rules an ancient civilization living in a gigantic cavern beneath the South Seas. The Dweller is the invention of the three Silent Ones, beings of immeasurable power. When they are forced to destroy their creation, which has turned evil, the immortal beings cry at their loss. Another Merritt novel, *The Metal Monster*, serialized in *Science & Invention* magazine in 1927, features living, thinking metal, more than a half-century before *Terminator 2: Judgment Day*.

In the 1930s, robots and robot brains were fairly common in science fiction, with both good and evil being represented. In nearly all cases, though, the robots are fully functional, thinking beings, not merely automatons obeying orders. In "Paradise and Iron" by Dr. Miles J. Breuer (*Amazing Stories Quarterly*, summer 1930), machines create but then later destroy an island paradise. In "Rex," by Harl Vincent, (*Astounding Stories*, June 1934), a super-robot surgeon nearly takes over the world but committs suicide because it can't experience emotions. Of course, Rex never realizes that the rage and the disappointment that drive it to destruction are emotions.

Edgar Rice Burroughs's 1939 magazine serial science-fiction novel *The Synthetic Men of Mars* (in hardcover from ERB Inc., 1940) features an army of androids created by a mad scientist, Ras Thavas, that he uses to try to conquer the independent city-states of Mars. Burroughs continued the story in the January 1941 issue of *Amazing Stories* with John Carter and the Giant of Mars. In that story, John Carter fights a student of Ras Thavas named Pew Mogel, who has created a 130-foot-tall warrior using the flesh and blood of great apes and Martians. Needless to say, John Carter destroys Pew Mogel, but even the Martian army is unable to defeat the giant, Joog. Finally, Carter persuades the monster to leave civilization alone, and it stalks off into the Martian desert.

Adventures in Outer Space

While these adventures were flourishing in *Argosy* and *Amazing Stories*, a new crew of space adventurers, including both an intelligent (though not particularly bright) robot and a clever android, were fighting villains in outer space in *Captain Future* magazine. The Captain Future novels were written by Edmond Hamilton and were very loosely based on an idea by the magazine editor Mort Weisinger. Captain Future, whose real name is Curt Newton, flies all over the solar system (and once or twice to the stars), battling enemies of the Planet Patrol and the Federation of Planets. Helping the captain in his mission are Grag, the seven-foot-tall robot; Otho, the android, an artificial, chemically created being; and Simon Wright, a human brain kept alive in a box. Together, the team of four is known as the Futuremen.

In the series, both Grag and Otho have distinct personalities. They aren't merely machines but thinking, rational beings. They are both human enough to have their own unusual pets. Though the two artificial Futuremen are not extremely bright, they fight the good fight with Captain Future and are able to think rationally and make important decisions when called upon to act. They are a form of artificial intelligence that helps mankind.

Sympathetic robots with artificial intelligence also appear in the stories "Helen O'Loy" by Lester del Rey (*Astounding Stories*, December 1938) and *I, Robot*, by Eando Binder (*Amazing Stories*, January 1939). (Isaac Asimov also used the title *I, Robot* for the first book in his famous Robot series.) In the first story, Helen is a female android whose creator falls in love with her. The narrator of the second story is Adam Link, who tells of his creation as the first thinking robot in autobiographical fashion and whose adventures continue for nearly a dozen episodes. In "Jay Score," by Eric Frank Russell (*Astounding Stories*, May 1941), the heroic pilot of a spaceship that nearly flies into the sun is revealed at the end of the story to be an android.

Isaac Asimov, the prolific science-fiction writer whose first work appeared in 1939, wrote numerous stories involving robots, of varying degrees of intelligence, throughout the 1940s and the 1950s. While most of Asimov's robot stories are puzzles in logic, sometimes they feature robots in humorous situations that they find difficult to comprehend.

In "Victory Unintentional," three specially designed robots are sent to negotiate with powerful beings that live on the heavy-gravity world Jupiter. Though the aliens are quite powerful, the superstrong robots are even more powerful and unwittingly reveal their incredible powers during their sojourn on the surface of the planet. When it comes time to negotiate, the once-belligerent beings are incredibly cooperative. It's not until afterward that the robots realize that the Jupiter aliens have mistaken the robots for ordinary humans and have been frightened into peace.

The Menace of Mechanical Life-Forms

In stark contrast to the Captain Future novels, Robert Bloch, the author of *Psycho*, wrote "It Happened Tomorrow" for *Astonishing Stories*, February 1943. The grim short novel tells of a revolt of the machines, very similar in theme and tone to King's short story version of "Trucks." In "It Happened Tomorrow," however, Bloch follows the events from the very beginning of the machine revolt to the very end, with the end being the total elimination of the human race by the bloodthirsty machine horde. As was the case in "Trucks," Bloch offers no explanation for the machines becoming intelligent, other than a few rambling sentences about the spark of life that once pushed man forward is now doing the same for inanimate objects.

Jack Williamson's short novel *With Folded Hands . . . (Astounding Stories*, May 1947) was the first science-fiction story to suggest that giving machines the ability to reason might be dangerous. In this provocative adventure, robots are created by a scientist who gives them one mission: to protect humans from harm. Since everything humans do involves some sort of risk, either physical or mental, the robots force people to do nothing. People are only able to sit with folded hands, while being watched and supervised by their robot guardians.

In the 1948 sequel to the book, . . . *And Searching Mind*, a small group of people revolt against the robots but eventually fail. The solution the rebels discover to doing nothing is to develop extraordinary mental powers. It's an answer that Williamson, in interviews years later, admits he never liked.[4]

Supercomputers Cause Mayhem

By the late 1940s, gigantic computing machines like ENIAC started making their presence felt in the scientific community and, shortly thereafter, became stock creations in science fiction. These early computers were the size of small buildings, and they devoured electricity like a kid in a candy store. While giant brains had appeared in older sci-fi stories, now they had a basis in reality. "A Logic Named Joe" by Murray Leinster (*Astounding Stories*, 1946) features a worldwide computer network that causes world-threatening chaos when it crashes. More typical of the time was "The Brain," a short novel by Alexander Blade (*Amazing Stories*, October 1948), in which a huge military thinking machine in the desert plots to control the world until it is destroyed from within by seemingly helpless prairie dogs nibbling at its circuits. In this story, not only are computers intelligent, they are also ambitious.

Charming Androids

In the late 1940s and from the 1950s on, the definitions of robots and androids began to merge, with many androids merely being robots with humanlike features. The character Data, on *Star Trek: The Next Generation*, is called an android, although he is shown to have mechanical insides. The terminator robots in the Terminator movie series are merely disguised to look like humans, whereas the replicants in *Blade Runner*, the film version of *Do Androids Dream of Electric Sheep?* are artificial people.

Asimov wrote two extremely popular novels, *The Caves of Steel*

(1952) and *The Naked Sun* (1955), featuring the robot detective R. Daneel Olivaw, in which the android teams up with a human detective to solve two important homicides. A sentient computer, which takes the form of a person to interact with its passengers, pilots the ultimate spaceship in Donald Wollheim's space adventure *Across Time* (1956). Thinking machines play an equally important role in the Berserker series by Fred Saberhagen. These series of novels and short stories began in the late 1960s. The Berserkers are killing machines left behind by a long-dead alien race, and they are programmed to hunt and destroy all life in the galaxy. The main theme of nearly all the Berserker stories is how men are able to out-think and thus outwit the artificially intelligent machines.

Ruling Their Masters

As computers grew increasingly sophisticated in the real world, they became more evident in our imaginary ones. On the TV series *Star Trek*, the starship *Enterprise* relies on a massive computer network to help run the ship, and the crew members carry tricorders, or miniature computers, with them on all their missions. Computers with artificial intelligence are the villains in a number of *Star Trek* episodes, including Landru, from "The Return of the Archons," and M5, in "The Ultimate Computer."[5]

In the film *2001: a Space Odyssey*, the most memorable character is the intelligent spaceship computer HAL 9000, which is willing to sacrifice its crew to complete its secret mission. The villain in *Colossus: The Forbin Project* is the worldwide defense net computer Colossus, as is Skynet, the supercomputer in the Terminator movie series. In all these examples, the artificial machine intelligence decides that it is smarter than humankind and thus destined to rule Earth, if not the universe.

A basic theme in science fiction during the last half-century has been that bigger means more complex—and more complex means more intelligent. Thus, it is no shock that in Robert A. Heinlein's *The Moon Is a Harsh Mistress* (1966), the computer system in charge of the moon colony develops a personality that the hero dubs Mycroft Holmes. Nor is it surprising that all of Earth's computer systems band together to secretly put a man on Mars in Frederik Pohl's novel *Man Plus* (1976). In the Pohl novel, the machines are primarily interested in guaranteeing that machine intelligence will not be wiped out by some cosmic disaster. Humans merely serve as a means to get the computer system to another planet. That such a network would not want to be discovered is the basis for John Varley's cautionary AI tale "Press Enter" (*Isaac Asimov's Science Fiction Magazine*, May 1984).

Computers in Modern Life

At the beginning of the twenty-first century, we live in a world where computers have crept into every facet of our lives. There are computers in our cars, in our watches, in our homes, in our hospitals, and in our supermarkets. Computers are everywhere. Still, most of these computers are not thinking machines. At present, they are merely highly advanced tools, capable of doing only what they are asked to do. They don't possess the artificial intelligence of the programs in *The Matrix*, of Isaac Asimov's robots, or of Data in *Star Trek: The Next Generation*. In its purest sense, artificial intelligence is still not yet possible. Despite all of our fictions, we have yet to come even close to creating an independent thinking machine. The world of "Trucks" is still only a TV movie.

That's not from lack of trying. One of the greatest challenges facing modern science is the creation of an artificially intelligent machine. The search for artificial intelligence is being conducted in labs all across the globe. It might not happen for another ten or twenty years, or it might take place tomorrow. No one is sure. But it will happen, and no doubt, the development of pure artificial intelligence will change the world in ways we cannot imagine, change it in ways that even the best writers of science-fiction and horror literature can't guess. But, like Stephen King with "Trucks," they will keep trying.

In the next section, we look into the science of artificial intelligence, where it has been, where it is going, and how it applies to cars today and in the near future. Are intelligent trucks part of our future? If so, will they be friendly or hostile to their creators?

Creating Artificial Intelligence

By definition, artificial intelligence is the ability of an artificial system, assumed in most cases to be a computer, to think independently. Or, as we stated earlier in this chapter, artificial intelligence is intelligence demonstrated by a nonliving being. The problem with such a definition is what exactly do we mean by intelligence? Computer intelligence has always been an amalgam of how we think and the way we program a machine to think.

Cybernetics

Norbert Wiener, one of the great intellects of the twentieth century, was among the first scientists to notice the similarities between human thought and machine operation. Wiener was the father of cybernetics, the study of "teleological mechanisms." The word *cybernetics* was invented by Wiener, who used it throughout his groundbreaking book *Cybernetics, or Control and Communication in the Animal and Machine* (1948). *Cybernetics* was derived from the Greek term *kubernetes*, meaning "steersman" or "pilot." The basic concepts of cybernetics came from the melding of electronic network theory, logic models, and neurology during World War II. Cybernetics evolved into the study of communication and control in humans and machines and between the two, usually involving regulatory feedback. Emphasis was focused on the relationships between the different parts of a system, rather than on the parts themselves. Cybernetics was a popular concept from the late 1940s into the 1960s, but it fell from popularity in the 1970s as the study of artificial intelligence went in new and totally different directions.

One of the main tenets of cybernetics involves negative feedback. Normally, a ship's pilot steers his boat in a fixed direction, toward a star or a lighthouse or along a certain compass heading. Whenever the wind or the sea throws the vessel off this heading, the helmsman brings it back on course. This process, in which deviations result in corrections back to a set point, is defined as negative feedback. Positive feedback occurs when deviations from a set point result in even further deviations. A room thermostat is an example of a machine that uses negative feedback. A thermostat measures the temperature of a room and turns the heat on or off to keep the room at a specific temperature.

Norbert Wiener felt that all intelligent behavior could be traced to negative feedback actions. Since feedback actions were expressible as algorithms, a procedure of well-defined instructions for getting something done, Wiener concluded that intelligent machines could be constructed merely by using algorithms.

Early Computer Science Theory

This simple method of looking at human logic and applying it to machines provided an early basis for computer science theory. The earliest attempts at creating artificial intelligence relied on reducing human thought processes to strictly logical steps and then turning these steps into code to be manipulated by a computer.

A computer functions at its lowest level by switching between two states: the binary number one for true and the binary representation for zero as false. Circuits are constructed from combinations of ones and zeros. This basic fact about circuits, however, carries some limitations. Computers can calculate only through long chains of yes-no, true-false statements of the type "If A is true, go to step B; if A is false, go to step C." In this type of setting, statements have to be entirely true or entirely false.

A statement that is true only 60 percent of the time is much more difficult to deal with. Ambiguity, mistakes, and partial amounts of information cause major problems for a computer that is best equipped to deal with the cut-and-dried, yes-no world of mathematics.

It took computer scientists years to realize that there was a near-insurmountable gap between the real world and the world as expressed in mathematical symbols. Binary logic was primarily suited for dealing with algebraic problems that could be expressed in terms of ones and zeros. Geometric and spatial problems were much more difficult. Situations where a symbol could have more than one meaning produced frequent mistakes.

Top-Down and Bottom-Up Approaches

This original type of artificial intelligence became known as the top-down approach. It used the combination if-then strings to apply intelligence to computers. It was very methodical and worked best in logic problems. The 1950s was the decade of top-down AI. Herbert Simon, an economist, and Allen Newell, a physicist and mathematician, designed a top-down program during this period known as Logic Theorist. Logic Theorist used decision trees to make its way down various branches until it arrived at a correct or incorrect answer. A typical decision tree for checking computer diagnostics was usually twenty to thirty pages long of yes-no answers. Top-down researchers tried to construct "rules of thought" through symbols. They then tried to combine these rules into new, meaningful ideas. This method of GPS (generalized problem solver) was typical of the top-down approach of the 1960s. For top-down researchers, the structure of the human brain was unimportant to their studies.

In 1956, Dartmouth College hosted a conference that launched the revolution in modern artificial intelligence research. The conference was organized by John McCarthy, the scientist who first coined the phrase *artificial intelligence*. Other computer theorists in attendance included Herbert Simon, Allen Newell, and Marvin Minsky. One of the main topics discussed at this conference was whether artificial intelligence was top-down, as was believed for years, or instead might be bottom-up.

Bottom-up is the belief that along with if-then, yes-no thinking, which is known as deductive reasoning, computers could also use inductive reasoning. Inductive reasoning is the process that studies information and, from this, derives a conclusion. Humans use both deductive and inductive reasoning.

The main problem with top-down, deductive reasoning, many scientists felt, was that it required a huge database of information to keep all of the possible yes-no, if-then, true-false facts available that a computer needed to solve a complicated problem. In the 1950s, computer speed was much more of a worry than it is today. It took a vast number of operations done over a number of days, winding through mazes and mazes of logic circuits, to arrive at a conclusion using top-down logic.

A typical top-down AI program is a computer chess program. Chess has a specific set of rules. Each move is well defined, and every piece is located at a specific place. Deep Blue, the computer program that beat world chess champion Gary Kasparov, does not possess artificial intelligence, but instead it uses a sophisticated if-then program that operates at lightning speeds to analyze its opponent's every move.

Top-down AI uses symbolic methods to mimic human logic; however, except for the endings of mystery novels where the inspector explains it all with a step-by-step analysis of each part of the crime, top-down logic is not the way most people think. In the human brain, numerous thoughts blaze through the mind all at the same time. Equally important, real life is not top-down. We rarely have complete information about every event in life. Nor are there rules governing what will happen next in regard to human behavior. Humans use common sense, intuition, and emotions, as well as logic, to reach conclusions. Programming these traits into a computer is impossible using top-down statements.

Bottom-up researchers feel that it is absolutely impossible to program computers with explicit rules for every possible situation. Such programming would require near infinite time, near infinite memory, and infinite patience. Instead, these researchers feel it's much more logical to give a computer a foundation of built-in capabilities for learning and then let the machine learn how to gather facts and draw conclusions. Bottom-up AI learns from what it does, devises its own rules, creates its own data, and then makes its own conclusions. It changes, adapts, and grows based on the network in which it lives. To develop true artificial intelligence, bottom-up scientists feel that a computer has to develop much as a human brain does, leading us to ask the obvious question: how are human brains and computer brains alike? And how are they different?

Brain versus Computer

The human brain and a computer are similar in that they both can do a number of things at once. A person's brain can calculate bills, daydream about vacations, think about when the boss will come to take that person to lunch, wonder what he or she will eat for lunch, and much more. A computer can typeset a book, print the pages of a book, download files from the Internet, save the pages it printed, calculate invoices for books, and do many other jobs. Humans and computers are both quite good at multitasking; however, a human brain thinks, whereas a computer does not.

A human brain receives information from the eyes, the ears, and the skin. A computer receives data from a keyboard, voice instructions, or data feeds. The human brain issues output to the eyes, the ears, and the skin, while the computer outputs to the screen, data feeds, and networks.

Human brains and computers are both quite complex. They both have hardware and software components, although they are entirely different in composition and materials. At present, we can build a computer. We can't yet build a human brain.

The most basic circuits of a computer rely on the true-false, on-off action of micro-switches. Neurons in our brain have similar states: excited and inhibited. When the voltage across a tissue membrane rises sharply, a neuron is excited and releases chemicals known as neurotransmitters that latch onto receptors of other neurons. When the voltage drops sharply, the neuron is inhibited. It's similar to the binary on-off states of a digital computer—or so it seems at first.

In reality, neural processes are much different from computer processes. Neurons behave in an analog, rather than a digital, fashion. A digital signal has two distinct voltage levels. An analog signal varies continually between a minimum and a maximum voltage. Events leading to neural excitement build up, as if climbing a hill. Moreover, ions may cross the cell membrane even if neurotransmitters aren't received. These ions may excite the neuron anyway. Sometimes a neuron oscillates between intense and minor excitement levels without any outside stimulation. The more a neuron excites itself, the more likely it will be susceptible to outside stimulation.

In a computer, the shape of the motherboard—large rectangle, small rectangle, oblong, oval, whatever—has no effect on how the computer works. Positioning components close together, shortening the circuit travel, and the choice of the actual components are conditions that certainly affect the processing speed and the power of the computer. Most motherboards are rectangles, however, and the actual shape really doesn't have a radical influence, such as popping an on switch to off or making an NOR into an XOR.

The neuron, though, is very different. There are approximately fifty neuron shapes that can change the state of the neuron from excited to inhibited or vice versa. For example, an incoming signal becomes weaker as it transverses a really long dendrite to the neuron body. A signal that travels along a short dendrite will be much more powerful when it hits the neuron body. In addition, it takes a higher dose of neurotransmitter to excite the big neuron than it does the small one.

Also, the brain uses a finite set of neurons to perform a flexible number of tasks in parallel. Neurons may interact in overlapping, multiple networks with the brain; a single neuron simultaneously communicates with many others in many neural networks. By intercommunicating constantly across these multiple networks, neurons learn to adapt and respond to their environments. Scientists think of the brain as a muscle: the more it is used, the stronger it becomes.

Artificial Life

The big question, of course, is how can we build such properties into a computer? The ultimate answer to this question is for scientists to produce bottom-up AI, which is known as "a-life," or "artificial life." With this type of computer intelligence, digital entities, nodes, and units not only adapt to their environment but reproduce, feed, and compete for resources. Their offspring evolve over generations to become better suited for their environment. While a-life

is mostly theoretical, certain digital creatures already exist in pro-
totype form.

Some a-life creatures are based on genetic algorithms that
affect their life expectancies. The creatures have genomes to define
what they're like, how they act, and what they do. To reproduce,
a-life cross breeds, and sometimes, as in the case of biological
life, the genomes are accidentally mutated by this action, creating
a new digital generation that is very different from its parents.
Our 1999 computer thriller *The Termination Node* featured mutating a-life executing a $50 billion Internet bank robbery.

Some a-life creatures grow through what might be considered
digital embryonics. Such beings exist in silicon, which is divided
into cells—where rows and columns intersect as on a sheet of
graph paper. Each cell contains a genome that's defined in random
access memory. At the beginning of its life, the creature is the
only individual in the silicon environment. This organism has a
certain number of cells, just like a human baby. Each cell has a
special function, although the creature can have many cells that do
the same thing. Which genes of the digital organism's cell will be
functional depends on the cell's row and column—its location—in
the creature.

When the a-life world begins, only one cell contains the entire
genome of the organism. This first cell divides just as it would in a
biological embryo. Now there are two digital cells, each with the
entire genome of the organism. Soon, the entire digital creature
exists, born digitally in a manner based on biology. By combining
digital embryonics with evolutionary algorithms, we have the
potential to grow truly complex a-life environments.

A-life is still in the experimental stage, but artificial intelligence
is already built into many of today's robots. In 1969, a robot called
Shakey was able to move around seven rooms that contained obstacles made of varying geometric shapes. Shakey received commands—
such as "Bring me a box"—from a computer console. Shakey rode
around on its wheels, avoided obstacles, snaked through the rooms,
and scooped up the box and returned it to some central location.
Shakey is the great-great-great-ancestor of the monster semi-trailer

rigs in "Trucks"—though where the trucks store their brains is still an open question.

How Plausible Is the Plot of "Trucks"?

Returning to the story and considering what we've just discussed, it is fairly easy to deduce that the monster machines attacking the diner and every place else on Earth are bottom-up thinking devices. There is no possible way the machines could have learned to murder people so efficiently without planning things in advance. The mere thought "kill" is useless if the methods of killing make little sense or if there is no way to find victims.

At the same time, the trucks also had to be at least partly top-down thinkers. Too much of their knowledge, from understanding how they are fueled and knowing humans' vulnerabilities, to knowing Morse Code, is learned information. Such information is not just present but has to be somehow fed into the minds of the trucks. Thus, we are forced to conclude that the killer trucks in "Trucks" and the movies based on the story are both bottom-up and top-down thinkers—meaning that they are doubly dangerous.

Before we abandon our helpless humans to the mercy of the killer trucks, there's one more topic worth discussing. The small number of humans left alive aren't rocket scientists. They have no idea how to fight the machines or even how to resist them. The entire story consists of the humans complaining that the events can't be happening and bemoaning their fate. There's no attempt to fight back, and it's made clear that they feel that any attempt to flee from the machines would result in disaster. In the end, we are left with a picture of desolation in which man is giving up mastery of the Earth without a fight. It's a depressing look at humanity's future. Maybe "Trucks" is the way things would happen—but, then again, maybe not.

In the story, the machines are huge and powerful, move quickly, and are lethal. Still, are they really as smart as the humans trapped in the diner think they are? Or are they actually morons with powerful and deadly tools? Even if the trucks had computer minds, would that necessarily make them more intelligent than ordinary humans? Or would it merely mean they were just able to solve math problems faster? Let's compare.

Many scientists believe that the human brain contains approximately 100 billion to 200 billion neurons that fire about 10 million billion times per second. Each neuron connects to roughly 10,000 other neurons. This is how the brain manages to handle trillions of operations in a second. It's an extremely complex neural network.

A computer-neural network is an extremely simplified version of a biological neural network. In the biological form, a neuron accepts input from its dendrites and supplies output to other neurons through its axons. The neuron applies weight to the connections, or synapses between dendrites and axons. A higher weight might be applied to a synapse related to touching fire than to a synapse about seeing the pretty orange color of a fireball.

Turning to the computerized version, each "input neuron" feeds information into every neuron in what is called the hidden layer, which may have one or multiple layers of neurons. If the hidden layer has two layers of neurons, for example, then every neuron in the first hidden layer feeds into every neuron in the second. Every neuron in the last hidden layer feeds into neurons in the output layer.

This design of a neural net provides different weights for the connections among neurons. While the brain receives input from many sources, such as the sensations on the skin, what we hear, what we smell, and so on, an artificial neural network takes input only from values we provide, and then it weighs everything and comes up with a best-guess answer.

We know, for example, that 1.00 + 1.00 = 2.00. An artificial neural net may not find that problem so easy. It may guess that the answer is 1.98 or 2.04. But the artificial brain will do quite well in guessing between a nerf soccer ball and a real soccer ball. Both are spheres. Both are the same size. One is soft—the nerf; one is hard—the soccer ball.

Various methods exist for applying weights to artificial neurons and for assembling the input, the hidden, and the output layers into network architectures. A neural network learns by adjusting the weights given to its neurons. A very common neural net architecture is called back propagation, which compares forecasts to actual events, then adjusts the weighted interconnections among neurons. Over time, as it compares more forecasts to actual events, the neural weights become more accurate. In a sense, the neural net learns and adjusts to its environment. So, neural nets can approximate human brains in some limited ways, but they are still far behind in other matters—like sight.

Let's assume for an instant that some sort of alien neural net intelligence floated down from the sky and gained control of our trucks. We still wonder whether they would have been able to wipe out civilization so quickly and efficiently as the story and the two movies seemed to imply. Controlling the mechanisms of a big truck means nothing if you don't know how to drive or if you can't see, and duplicating the human brain's mechanism for sight is not feasible for most machines.

Vision is one of the five senses that is managed easily by the human brain but is extremely difficult to program for an artificial intelligence. What we do naturally is nearly impossible to duplicate with a computer. The sensory abilities and the neural architecture of animals, including people, is based on what they need. Important factors include reaction time, size, and the complexity of the objects that are recognized by the brain.

Human visual perception is based on incredibly complex input. The retina uses cells called rods to handle incoming light and other cells called cones to handle incoming color. With approximately 100 million rods and cones, the retina processes images at the rate of 10 billion calculations per second.

After image preprocessing by the retina, the cerebral cortex of the brain takes over. Vision centers for this purpose occupy more than half of the cortex. At this point, the brain hasn't even begun to determine what the person is looking at—a flower, a field, a crowd of people—much less fit the objects into a moving scene and analyze who and what the person knows or how he or she plans to react. All the brain has is raw image information. It's a vast collection of individual bits of data that needs to be processed, which is exactly what the human mind does so well.

Our visual skills are far better than the world's best cameras. We instantly recognize and respond to textures, lights, and shades. Our brains create an imprint that is complete with texture, object boundaries, and compensation for clouds, brightness, and shadow. In dim or flickering light, we can still recognize our friends and our enemies. The outline of a person's profile in the darkness is often-times enough for visual recognition. We can even interpret and process strange anomalies, such as optical illusions. Computers can't see optical illusions, much less explain them. More to the point, computers cannot come close to duplicating human vision, nor does that possibility seem likely any time in the near future. Machine vision relies on physics, neurobiology, signal processing, and spatial geometry. Constructing a computer with vision comparable to a human eye is a dream of modern computer science and artificial intelligence. But until that and other problems involving artificial intelligence are solved, trucks won't be running free across our highways.

The Cars and Trucks of Tomorrow

In the late 1940s and early 1950s, science sections of the Sunday newspaper were filled with stories of mini-helicopters flying com-muters from the suburbs to work. High-speed bullet trains filled cartoons with clean, inexpensive transportation for people who couldn't afford a family helicopter. Needless to say, no one ever dis-cussed monitoring traffic in the air lanes, nor did anyone seem to worry about the huge cost of monorails. It was the future, and everyone knew the future took care of itself.

Fifty years later, at the beginning of the twenty-first century, we're still looking for those one-man helicopters. Traffic congestion on the major highways that feed into our big cities is at an all-time high. Rail transportation is underfunded and by no means univer-sal or cheap. Except for slightly better gas mileage—and it's only slightly, due to the gas-guzzling SUV monsters so many people drive—the world of automobile transportation in the United States is little better than it was a half-century ago. Hybrids are in their earliest incarnations, with slightly improved gas mileage, but we're still reading articles in the science sections of the Sunday paper promising miracle cars in the near future.

Computers are everywhere in our modern society, and the smaller they are built the more difficult the type of problems they can manage. Most modern cars already contain a huge number of computer switches and circuit boards that run the motor and accessories at top efficiency. Auto designers, working with engineers, are adding layer upon layer of computerized devices to our cars, trying to modify and change our driving experiences from daily nightmares into relaxed, comfortable excursions. Plus, they're working to integrate the car and the road into one system that offers extra protection for the user.

VII (Vehicle-Infrastructure Integration) is a U.S. Department of Transportation initiative that involves auto makers; federal, state, and city governments; technology companies; and trade associations. Working together, they are looking to develop high standards for technology and then deploy this new technology to the nation's highway system.

"We are taking advantage of all reasonable means to prevent crashes and reduce the deaths on our highways," said U.S. transportation secretary Norman Y. Mineta in 2004. "Treating roads as an extension of vehicles, using both design and technology, will help prevent crashes and make driving safer."[6]

New cars today come equipped with nearly two hundred sensors that measure everything from engine processes to outside air temperature. Scientists hope to use that information to run cars more efficiently and to safely manage transportation systems. "Recently we've begun to explore vehicle-to-roadside and vehicle-to-vehicle communication," a U.S. Federal Highway Administration (FHA) official told *National Geographic News*. "That's sort of a hot new area that offers potential for a whole new family of services."[7]

Ford Motor Company's Smart, Safe Research Vehicle is an

enhanced version of the company's Explorer model. The concept automobile includes such "intelligent" safety features as a navigation system that uses real-time traffic information to suggest alternate routes around congested areas.[8]

The European Union is working extremely hard on what is known as the eCall system, in which vehicles automatically contact emergency services in the event of an accident. This device could cut fatalities by 5 to 15 percent if all new cars were equipped with it by 2010, according to Rosalie Zobel, a director of the European Union's eCall Commission's information society's directorate-general.[9]

Devices such as adaptive cruise control, which help to prevent rear-end collisions by monitoring the position of vehicles in front of the car, could stop 4,000 accidents a year even if only 3 percent of cars had the technology installed by 2010, the official explained. Other devices that monitor a vehicle's lane position could prevent 1,500 accidents a year if only 0.6 percent of vehicles had the products by 2010, the commission said. A technology that monitors drivers' eye movement and triggers alarms when they get sleepy could help to prevent 30 percent of all fatal motorway crashes and 9 percent of all fatal accidents.[10]

It's not just the trucks that are getting more intelligent; the cars are, too, and we're helping them in every way we can. We can only hope that in making our vehicles more intelligent, we don't open a Pandora's box and make them more intelligent than humans.

Friend or Foe

On the surface, the basic story of "Trucks" is a simple one. Somehow, trucks and other mechanical vehicles develop artificial intelligence and start to kill human beings. There are few subtleties, and the human interaction is not really a major focus of the story. It's a horror story, first and foremost, and it manages to paint a convincing picture of the complete collapse of human civilization in a matter of mere days. Our modern world is totally dependent on automobiles. Without them, we are helpless. Plus, there are so many trucks and cars in the United States that our chance of surviving against them in any sort of actual war is practically nil.

Artificial intelligence in cars and trucks isn't here yet and most likely won't be for some years to come. But it's coming sooner than just about anyone thinks, and by the time this book is in its fifteenth or so printing, artificially intelligent vehicles will be here. It won't happen all at once, but, as we've noted in previous sections, it's already begun. Cars are learning to monitor themselves and keep track of when they need service and exactly what for. Huge transportation networks are already available so that we can tell our vehicles where we want to go and be guided there by a voice in the dashboard, taking the fastest route, the most scenic route, or the one with the most rest stops. The time is fast approaching when someone will open his or her car door and the vehicle will drive the person to work without having to be told. Then it will come and pick its owner up after checking out of the parking garage where it's been playing poker the whole day with three other cars and an AI street-sweeper. It all sounds like science fiction now, but fifty years ago, cruise control, GPS (global positioning system), and heated seats sounded equally impossible.

If intelligent vehicles are possible, then what can we do to make sure something like "Trucks" doesn't happen? Is there some action we could take today to ensure that the nightmarish scenario King describes in his story never occurs? Is there anything we can do other than abandon all of our cars and trucks and return to walking? Or forbid all research into artificial intelligence? Isn't there a solution that's a bit more practical and a lot less draconian? Of course there is: we can design our artificial intelligence to be friendly.

Looking at the concept of artificial intelligence logically, we realize that AI can be friendly to humans, hostile to humans, or strictly neutral. We'll consider neutral to be pretty much the same as friendly because, in effect, that's exactly what such actions would be. In "Trucks," the AI controlling the machinery is definitely hostile to humans. As pointed out earlier in this chapter, when one of the characters is asked why he thinks that machines have come to life, the lead character states, "Maybe they're mad." That's about as close a reason given for the trucks' motivation and actions in the story. In the two movie versions, machines are brought to life by outlandish explanations involving Area 51, flying saucers, and maybe a comet. Nowhere is there any explanation of why trucks would be hostile to humans. After all, we created them. Why are they so mad?

As is pointed out on the Singularity Institute for Artificial Intelligence Web site, "Humans are uniquely ill-suited to solving problems in AI."[11] That's because when we think of alien (or, in this case, artificial) life-forms, we immediately assume that they will think and act exactly like us. But we are products of our environment and our heredity. We are influenced by thousands of years of history and, before that, a million or so years of evolution.

We tend to consider anything new and different to be hostile. In movies like *The Matrix* or *The Terminator*, no one watching the films objects to the fact that incredibly powerful thinking machines—as soon as they are given the chance—immediately try to destroy the human race. There is no attempt to bargain, negotiate, or even discuss terms. It's just attack and kill, with a single-mindedness of machines that have been set up to run a program from beginning to end without any glitches in the activities.

Such notions, however, are based on the idea that artificial intelligence would act with the same belligerence as a typical human being. Fortunately, there's no evidence to support such a view. According to the Singularity Institute, "A real AI wouldn't be a computer program any more than a human is an amoeba; most of the complexity would be as far from the program level as a human's complexity is distant from the cellular level."[12] Machines do not know that they are machines. Most of them are built for a specific job, and they do that job well. They are not self-aware. A true artificial intelligence would not only realize that it is a machine, created to perform a certain task, but could also have access to its source code—which is an interesting trait not possessed by humans.

Another common theme in stories featuring AI is that the behavior of beings with AI is dictated by force. They are made to act the way they do, as slaves are. They are given orders to obey, which forces them to do things they would never do otherwise.

Again, this notion makes the assumption that AI beings are just like people and act exactly like people. There is no logical reason that an intelligently designed artificial intelligence would be built with all sorts of rules forcing it to do certain jobs and not to do certain other jobs. Why create a machine to be your enemy when it's much easier, and much more logical, to program it to be your friend?

Victims of Possession

Examining "Trucks" with the concept of friendly artificial intelligence firmly in place, we are forced to conclude that the story as told makes little sense if we assume that the vehicles in question suddenly just come alive. There's no believable explanation why trucks gifted with artificial intelligence would want to kill their builders. This leads us to conclude that the machines in "Trucks" are not the product of artificial intelligence but instead are the victims of possession. That somehow, from somewhere, a horde of alien minds, millions upon millions of them, have descended onto Earth and mentally taken control of all the cars and the trucks on the planet. "Trucks" isn't a story of artificial intelligence but instead is an account of an alien invasion—which just happens to be the subject of our next chapter.

3

THEY CAME FROM OUTER SPACE

Dreamcatcher • *The Tommyknockers*

> If Gard's Tommyknockers had appeared to Beach in person,
> carrying nuclear weapons and proposing that he plant one
> in each of the world's seven largest cities, Beach would
> have immediately started phoning for plane tickets.
>
> —*The Tommyknockers*

A prominent theme in Stephen King's books is the threat of alien invasion. The aliens tend to be harmful, if not evil, and come in many forms, including fungi and ectoplasmic blue goo.

Catching Dreams and Knocking Tommies

A common science-fiction topic is that there is life on other planets and that aliens from these other worlds come to Earth and attack us. This chapter explores the alien-attack theme in novels such as *Dreamcatcher* and *Tommyknockers* and discusses whether alien life is possible, why aliens might be hostile toward us (or not), and why we haven't had alien visitors yet, at least that we know of.

First, let's take a look at *Dreamcatcher* (2001). You might remember that Stephen King was hit by a van in 1999, and he nearly died. His first novel after the accident was *Dreamcatcher*, which is possibly the most graphically violent book of his career.

In this novel, an alien spaceship crashes in the woods outside of Derry, Maine, the setting of other King novels such as *It*, *Insomnia*, and *Bag of Bones*. The *Dreamcatcher* aliens bring a contagious, killer funguslike growth called the byrus to Earth. If the infection spreads all over the planet, it could mean the end of humanity.

The aliens cannot survive in Earth's atmosphere and in Maine's climate, and they also can't overpower the paramilitary forces that show up to destroy them. In the woods with the aliens are four men, who meet there each year to hunt and to rekindle their lifetime friendship. Each man has an extrasensory power that is rooted in his childhood in Derry.

When they were children, the four boys rescued a boy with Down syndrome, Duddits (his real name is Douglas), from high school bullies. Their saving of Duddits, and his gratitude, bound the four grown men in tight friendship. Duddits was their one, true heroic act, and it stays with them. Other than Duddits, their lives have been less than heroic. Henry Devlin is a suicidal psychother-apist, Peter Moore is an alcoholic car salesman who used to dream about working at NASA, Beaver Clarendon messed up his marriage and entire life and cusses about it constantly, and finally, Jonesy was in a near-fatal car accident, which reminds us of Stephen King's own near-fatal accident directly before he started writing *Dreamcatcher*.

One day while Henry and Peter are out getting groceries, Jonesy runs into a man named Richard McCarthy wandering in the woods and accidentally shoots him. McCarthy doesn't die and tells Jonesy that he's been lost in the woods for days. He has extreme digestive problems (including a lot of flatulence). A blizzard rages outside the hunting cabin, while inside the small enclosure, McCarthy's condition gets worse and worse, and Jonesy and Beaver must deal with whatever's wrong with him. And what's wrong with Richard McCarthy is pretty darned awful.

McCarthy has the alien byrus infection. Because the aliens cannot survive on Earth and are dying, as they decay, the byrus infection spreads to humans. Some of the human hosts obtain telepathic powers and red fungal growths; others die from gruesome Ebola-like conditions.

Similar to *Dreamcatchers* is *The Tommyknockers* (1987), in which a spaceship lands in the woods near Bobbi Anderson's home. The spaceship is vibrating like a tuning fork, and when she tries to scratch it with a screwdriver, nothing happens. The metal is greasy and is nothing like any metal on Earth. Reports of aliens and spaceships in the area are chalked up to air turbulence and other atmospheric events. Flying saucer interest has decreased, and the air force doesn't even pay attention to reports of spaceships.

Meanwhile, an alcoholic named Gardener, who is friends with Bobbi, starts hearing radio stations playing music inside his head. The noise is emanating from a place in his skull where doctors implanted a piece of metal. On top of this, Bobbi's water tank contains some sort of force field with a green mist and ectoplasmic blue goo in it.

If all that isn't weird enough, things get even stranger. Bobbi's Tomcat farm vehicle has an extra stick shift setting on it—for no reason, it's just there one day, and it's to engage the "up" position. Her farm vehicle can levitate.

Bobbi's typewriter becomes all powerful and cranks out a four-hundred-page novel for her in three weeks. Green light flows from the typewriter, and the keys press themselves down, writing the book.

Characteristic of King's novels, Bobbi is somehow using mental telepathy to write the book while she works in the garden and walks in the woods. Because she has mental telepathy, Bobbi is also able to discern that the mailman is cheating on his wife.

People are bleeding and going crazy. The flying saucer in the woods is still vibrating. Bobbi senses that the aliens in the ship are all dead. Her teeth and hair start falling out, and she drops far too much weight. And then the rest of the townspeople begin to lose their teeth and hair, and they drop weight, too. Bobbi guesses that the outer layer of the ship is oxidizing as the air hits it, and whatever results from the oxidation is making people bleed, go nuts, lose their hair, and drop weight. It's like nuclear fallout, she tells Gardener. Before long, the entire town has fallen prey to the alien fallout, nobody has teeth or hair, and everyone can read minds.

A weird air blows from the ship, and anyone who has been transformed must remain in town, close to the weird air—otherwise, they will die. Bobbi's skin turns into a semitransparent jelly, and her body grows short, thick, and alien. She begins to resemble the dead aliens, who have long snouts, scaly transparent skin, no teeth, and permanent snarls.

How wild is the notion of alien visitors, as in *Dreamcatcher* and *The Tommyknockers*? Is alien life necessarily hostile, and where are all these aliens, anyway? The possibility of intelligent life in other solar systems is one of the most hotly debated topics in modern astronomy. Let's examine each side of the argument and see what it tells us about aliens.

Invaders from Where?

The concept that we are not alone in the universe isn't new. It's a subject that's been examined by philosophers, as well as scientists, for thousands of years. Pluralism is defined as the belief that the universe is filled with planets harboring intelligent life.

Pluralism versus Plenitude

Pluralism was first championed by the Greek atomist philosophers Leucippus, Democritus of Abedera, and Epicurus in the fifth century B.C. These men believed that Earth was the product of a chance collision of indestructible particles known as atoms. Since one world had been formed in such a fashion, they argued that other worlds with intelligent life were possible as well. Opposing the atomists were Aristotle and Plato, who said that Earth was unique and no other worlds or intelligent life-forms existed.

Needless to say, the Aristotelian view of the universe was accepted by the Catholic Church because that viewpoint placed man in a special place in the universe. In the late twelfth century, however, a number of scholars raised some serious religious arguments against Aristotle's belief that there was only one possible cosmos. Because God was omnipotent, these men declared, stating that God created only one universe was in a sense placing restrictions on God's power, which would thus imply that God was not all-powerful. In 1277, the church eased its stance on the unique nature of the universe. Catholic doctrine was revised to say that God could have created other worlds with intelligent beings but didn't.

It was a very small step for science but a major one. Following the same line of reasoning, Nicholas of Cusa in 1440 declared that whatever God could do would be done, a belief that became known as plenitude. Less than a century later, Copernicus argued convincingly that the sun was the center of the solar system and Earth was merely a planet revolving about it. Copernicus wisely didn't delve into the theological ramifications of his discovery, but other scholars and philosophers were soon debating the possibility of life on other worlds.

For the next several hundred years, proponents and opponents of plenitude and pluralism argued about God's purpose in creating a universe filled with stars and planets. As usual, in debates where men tried to explain God's purpose, neither side convinced the other that they were correct. Fortunately, by the mid-nineteenth century, developments in science and astronomy made such debates moot. God's intent faded in the face of the theory of evolution, and a scientific view of the universe slowly but surely replaced the religious one.

Still, while pluralism and plenitude were interesting theories, no factual evidence existed to back up either philosophy. There were numerous theories about life on other planets, but nothing could be proven. Telescopes could show only so much. There were no canals on Mars, and the clouds of Venus didn't shroud gigantic oceans or primeval forests. The only aliens from other worlds appeared in science-fiction books and magazines or in comic books like *Superman*.

Cold War Angst and Paranoia

World War II left the American public cynical and disillusioned. The atomic bomb displayed the frightening power of new technology. The Yalta conference, concentration camps, and the Berlin blockade pushed distrust of politicians to an all-time high. Nuclear power plant stories filled newspaper Sunday supplements. Space travel seemed only a few years away. After years of stories about life on other planets, people started to wonder where the aliens were and whether it was possible that we were the most intelligent species in the universe. If the galaxy was so huge and full of life, why hadn't other life-forms contacted us? (Of course, flying saucer advocates claim that aliens have contacted us. For purposes of this discussion, we'll consider only verified alien landings.)

As always when a question is raised, someone was there with an answer. It's not surprising that the average citizen, living in a cold war atmosphere of distrust and misinformation, was more than willing to believe that our government was concealing the truth about aliens. In 1947, we suddenly learned from several nonfiction books and magazine articles that other eyes were watching. It was the beginning of the flying saucer craze. Flying saucers became part of our vocabulary. They were featured in innumerable magazine stories and tell-all books, and they dominated late-night talk-show radio. Saucers have remained in our skies for the last half-century, despite the lack of any conclusive evidence proving their existence. In 2001, surveys indicated that a majority of people in this country believe that Earth has been visited by aliens.

Suddenly, the question wasn't whether aliens existed on other-planets. Instead, the question became: why are the aliens spying on us?

Flying saucers were a major setback for scientists trying to prove that extraterrestrial life existed in the galaxy. Frank Drake, one of the leading astronomers of the twentieth century, put it best when he stated:

The problem is that no civilization can thrive on falsehood. In the end, false "knowledge" leads to wrong decisions, wrong choices of technologies, a wrong distribution of resources, wrong priorities, wrong choice of leaders. And civilization as a whole is the loser. A prime illustration of this is the distribution of resources invested in attempts to understand life in the universe. There is widespread public confusion as to the relative promise of pseudo-scientific studies of UFOs . . . as compared with true scientific programs to find life on other planets. . . .

The consequence is that far more attention is given to the pseudoscience than to the real science.[1]

So, whether you believe in flying saucers or think that they're an ongoing money-making hoax, where are the aliens? If they're here, why are they so shy? After all, building a ship that can navigate the far reaches of outer space takes a fairly sophisticated and advanced civilization—one a good deal more advanced than ours. They can't all be tongue-tied or hiding under sofa cushions. Surely one of them has something to say to the world at large. Even a quick "Hi" would satisfy most people.

Imagine the worldwide excitement that would erupt on Earth if a flying saucer landed on the White House lawn and a humanoid figure stepped out to bring greetings from another planet. See *The Day the Earth Stood Still*, one of the best science-fiction movies ever produced about this topic. It was filmed during a time when science-fiction films weren't made only to sell toys and feature great special effects. A visit like that would change the world overnight. We would actually know that we aren't alone in the universe, not to mention that we aren't the smartest or even the strongest kids in the neighborhood.

The Drake Equation

Pluralism and plenitude are interesting theories but have no basis in fact. Are there alien civilizations in the galaxy? Do we have any proof at all that we are not alone in the universe, other than religious doctrine?

In the 1950s, Drake proposed an equation to estimate the number of intelligent species in our galaxy, the Milky Way. This equation served as the rallying point for the earliest efforts to use radio telescopes to detect signals sent by other highly advanced civilizations. Run for months by Drake, Project Ozma had no success in detecting the all-important radio signals from other star systems; however, a far greater effort was organized by scientists and continues through this day. The Search for Extraterrestrial Intelligence (SETI) served as the background for Carl Sagan's book (later made into the movie) *Contact*.

The Drake equation is a fairly simple multiplication problem:

$$N = R^* \times f_p \times ne \times f_t \times f_i \times f_c \times L$$

where

N is the number of intelligent civilizations in the galaxy (the number we are looking for).

R^* is the birthrate of suitable stars for life in the Milky Way galaxy measured in stars per year.

f_p is the fraction of stars with planets.

ne is the number of planets in a star's habitable zone (which we define further on).

f_t is the percentage of civilizations that have the technology and the desire to communicate with other worlds.

f_i is the fraction of habitable planets where life does arise.

f_c is the fraction of planets inhabited by intelligent beings.

L is the average in years of how long the technologically advanced civilizations last. In other words, how long is it from the time aliens invent radios to when their civilization either destroys itself or disappears?

The only phrase that's confusing is a "star's habitable zone." In simplest terms, the phrase refers to the imaginary shell around a star where the surface temperature of a planet in that shell would be conducive to the origin and the development of life. As far as humanity is concerned, the habitable zone around a star is the space where planets that have water in liquid form exist, water being the most basic necessity for life. In our solar system, Earth is obviously in the habitable zone. Venus, which is too close to the sun, is not. Nor is Mars, which is too far away.

In the 1950s, when Drake invented the equation, many of the numbers and the fractions were not known. As our knowledge of astronomy grew, more of the numbers became available. Still, some were based more on hopes and beliefs than on actual information.

The Principle of Mediocrity

A very popular theory about the universe that is believed by Carl Sagan and other space scientists is known as the principle of mediocrity (sometimes called the Copernican principle). This theory, based entirely on logic, states that since Earth appears to be a quite typical and common planet, intelligence has a very high probability of emerging on any planet similar to Earth after 3.5 billion years of evolution. In simplest terms, the principle of mediocrity states that Earth isn't special, so there should be lots of other planets with life on them.

Belief in the principle of mediocrity fuels the scientists who believe in SETI. It is also what makes the Drake equation work. Without it, we'd most likely not have stories about visitors in flying saucers. In the last decade, however, a growing number of scientists have been studying the principle of mediocrity, and they find it wanting. We'll discuss this idea in a minute.

The Case for Intelligent Alien Civilizations

For the moment, let's plug some numbers into the Drake equation. To put things in perspective, let's use the numbers that Frank Drake and Carl Sagan used and see how many intelligent extraterrestrial civilizations are out there.

R^* has been estimated by astronomers to be between 1 and 10 stars per year. Drake picked 5 as an average.

For f_p, the number of stars with planets, Sagan believed that a majority of stars had planets. In the last few years, we've actually located some. Let's be somewhat conservative and pick 20 percent, or 1 out of every 5 stars.

For ne, the number of planets that exist in the habitable zone, if we use our solar system as a model (the principle of mediocrity), then the number is 1.

For f_i, the percentage of worlds like Earth where life begins, Drake and Sagan chose 100 percent, again using Earth as their model.

For f_c, the percentage of planets with intelligent life, SETI scientists argue that evolution over billions of years leads to intelligence, so again the percentage could also be 100 percent.

For f_t, intelligent species who develop the technology and the desire to communicate with other worlds, Drake estimates that this value is 100 percent.

The math is pretty basic. Multiply all the numbers we have so far, and (surprise, surprise) we end up with an equation that $N = L$. This is the same result that Drake and Sagan arrived at years ago: the number of intelligent civilizations in the galaxy equals the average lifetime of technologically advanced civilizations.

Again, let's assume that Earth is average (using the principle of mediocrity). If our civilization self-destructed next year due to terrorism or the release of a deadly plague virus, then L would be approximately 100, meaning that our galaxy would be home to a hundred alien civilizations. Considering that there are somewhere between 200 and 400 billion stars in our galaxy, we are suddenly faced with the possibility of one civilization per 2 billion to 4 billion stars. It's no wonder we haven't been contacted by aliens. Reducing it to more human terms, it would be as if two people were born on the Earth during the last fifty years, separated not only by time but by thousands of miles. Neither one has any clue about the other except that they have the same birthmarks. Then somehow, they must find each other, searching on foot.

Drake and Sagan both knew that L, the lifetime of a technologically advanced civilization, was the great stumbling block in the Drake equation; however, both men were not only scientists but optimists. Drake felt that a technological civilization might last for ten thousand years. Thus, he estimated that there were ten thousand advanced civilizations in the galaxy. Carl Sagan, who was even more of an optimistic than Drake was, estimated in 1974 that there might be a million civilized planets in our galaxy. This would leave us with one civilization per 20 to 40 million stars, still somewhat of a daunting search. Other scientists believe that number to be much too low. They estimate that there could be hundreds of thousands of such civilizations—which would mean we're not as alone as we thought.

More important to our concerns, the Drake equation, working with the figures cited, gives us estimates ranging from one hundred to ten thousand civilizations in the galaxy.

Our home galaxy, the Milky Way, is only one hundred thousand light-years across. There are approximately twenty other galaxies, some much larger than the Milky Way and some much smaller, within 3 million light-years, increasing our range of possible civilizations from two thousand to two hundred thousand.

Even if there are only a few hundred civilizations per galaxy, there are a lot of galaxies in the known universe. A recent estimate placed the number at 50 billion galaxies. Assuming one hundred civilizations per galaxy, that still results in 5,000 billion (5,000,000,000,000) intelligent civilizations in the known universe. If we take Drake's more optimistic guess, we're talking about 500,000,000,000,000 (500 trillion!) advanced civilizations in the universe.

But don't tell Peter D. Ward and Donald Brownlee that, because they've raised some serious doubts about the Drake equation.

Rare Earth?

In their 2000 book *Rare Earth: Why Complex Life Is Uncommon in the Universe*, Peter F. Ward and Donald Brownlee discuss the principle of mediocrity. The two scientists argue that perhaps we've gone too far in trying to prove that man isn't special. They argue that in our attempts to understand the universe surrounding us, we've downsized the significance of life on Earth. Perhaps, they propose, life is not common and the principle of mediocrity isn't true. Maybe, as the ancient religious thinkers believe, mankind is unique.

It's a startling proposition, but the two men build a compelling case. Chapter by chapter, they examine each factor in Drake's famous equation and arrive at totally different conclusions.

The Case against Intelligent Alien Civilizations

The basic problem with the Drake equation is that it's a series of numbers multiplied together that gives us a final answer. In any multiplication problem, if any one number is zero, the answer is zero. If any one number is a very small fraction, the answer becomes a small fraction. If the assumptions used to produce those numbers are incorrect, then the numbers are invalid. In the Drake equation, too many of the numbers are based entirely on speculation, hope, and faith, not on fact.

Let's examine four of the most troubling figures in the Drake equation. Instead of taking the optimistic viewpoint adopted by the people working on SETI, let's instead look at them with a much more pessimistic eye, the viewpoint of *Rare Earth*.

For example, f_p is the fraction of stars with planets. Our solar system has eight officially recognized planets. Carl Sagan argued that an average solar system most likely would have ten or more. Other noted scientists of the 1970s and 1980s felt that ten planets was a good estimate; however, major strides in astronomy during the last decades have caused astronomers to rethink this belief.

In the last ten years, scientists have discovered twenty-seven planetary bodies circling other stars. All of the planets we've located are huge, about the size of Jupiter, the largest planet in our solar system. Astronomers studied numerous stars to find the twenty-seven stars with huge planets circling them. There's no method yet developed to locate smaller, Earth-sized planets. So, the guess that f_p amounts to one of every five stars having planets could be extremely high. We probably won't have a good estimate on the average until we start traveling to other solar systems.

Ne is the number of planets in a star's habitable zone. Until recently, the habitable zone has been defined as the appropriate distance from a star that enables liquid water to exist and complex life to develop. Earth is the only planet in the habitable zone of our solar system. It's possible that some sort of simple biological life might exist or may once have existed on Mars, but that's not been proven.

Ward and Brownlee argue in their book that based on what we've learned about astronomy in the last few decades, it's clear that habitable zones are a lot more complicated than anything imagined by Drake and Sagan in the 1960s or 1970s. For example, they point out that the presence of Jupiter, a massive gas giant much farther out in our solar system, was a crucial factor in life developing on Earth. Jupiter's immense gravitational pull attracted most comets to it, instead of allowing them to crash into Earth. Without a Jupiter-sized planet serving as this type of shield against stray comets, life on Earth would have been subject to mass-extinction events and planetary disasters caused by space collisions.

Therefore, the habitable zone of a solar system isn't merely based on the location of a planet in a solar system, but on the location of other planets in the system as well. Which makes the existence of habitable zones a great deal less probable than was once considered.

Ward and Brownlee take habitable zones a step further by considering the zone of space where animal life, not merely biological life, could develop. Biological life, such as primitive bacteria, can exist in extreme heat or extreme cold. Humans can't, and the difference needs to be taken into account.

F_i is the fraction of habitable planets where life does arise. In *Rare Earth*, the authors examine the length of time, measured in billions of years, necessary for a habitable zone to exist in relative stability for evolution to take place. Using Earth as our model, that time zone needs to be at least 3 billion years long. They point out that our sun, a G2-type star, has a lifetime of 10 billion years, more than enough time for complex life to develop.

G2 suns, however, are not the most common stars in the galaxy. That honor belongs to M stars, which have a mass of only about 10 percent of our sun. As these stars don't emit nearly as much heat as

Sol, the habitable zone around them is much closer to the star itself. Planets would need to orbit much nearer to the sun, which leads to a host of problems. Gravitational tidal effects from the star lock the planet into an orbit where only one side of the planet faces the sun—an orbit like that of Mercury. It's an orbit that's not conducive to human life.

Going in the other direction, stars more massive than our sun have much shorter lifetimes. Sol is predicted to remain stable for 10 billion years. A star 50 percent more massive than our sun would last only 2 billion years before entering the red giant stage. When a regular star transforms into a red giant, all planets in the original habitable zone in space are burned away, as new habitable zones are established millions of miles farther out.

Big hot stars like Sirius also generate a lot of their energy as ultraviolet (UV) light. UV light is fatal to biological molecules, so any star system with a high-density sun wouldn't be the home of carbon-based beings. Thus f_i might be a complex problem, involving habitable zones, the structure of the solar system, and the type of suns at the center of the same system.

The situation becomes murkier and definitely not better when we factor in the next variable: f_i is the fraction of planets on which life develops intelligence. Drake felt that every place where life began, intelligent life would arise. That's an optimistic viewpoint, based entirely on the fact that intelligent life developed on Earth. The more we learn about the slow, complex path of evolution from single-celled organisms to a walking, talking, thinking man, the less sure that number becomes. Drake and Sagan argued that intelligence was inevitable on any planet where life began. Many scientists now believe that considering the more than 3 billion years it took for complex, intelligent life to evolve on Earth, we were very lucky.

If we could compress time so that one second equaled ten thousand years, all of humanity's recorded history could be squeezed into one second. Mankind's rise from simple predator to ruler of the earth fills three seconds; however, the time it took for one-celled organisms to evolve into intelligent life spans two and a half days. During that long, slow rise, paleontologists know of at least ten extinction-level events where more than half of all known life on Earth was destroyed. Optimists would argue that the development of intelligent life on Earth despite these ten extinction-level events demonstrates that complex life is inevitable. Pessimists would argue that we've been fortunate and the next extinction-level event could be our last one.

If f_i is less than 100 percent, then what of f_t, the number of alien races that will try to communicate with other species from another planet? Frank Drake and the scientists of SETI believe that percentage to be 100 percent—that every race of beings in the universe wants to discover intelligent life elsewhere. They base their assumption on our behavior. But aliens, being alien, probably will have little or nothing in common with us. They might not be curious, or a vast number of them might not want to use their resources to contact other races in space. Instead, they may spend their money on the poor, the homeless, and the hungry. Thus f_t could be 100 percent, but it just as likely could be 1 percent. If the Drake equation is to have any relevance, we need to consider both possibilities.

If we plug all the worst-case-scenario numbers into the Drake equation, then estimate that L is 1,000 rather than 10,000, the result is that there may be only one civilization in the Milky Way galaxy. Like it or not, we may actually be alone in our galaxy.

Friend or Foe?

Still, if intelligent life does exist elsewhere in the universe, we have absolutely no way of knowing whether it would be friendly or hostile to us, or whether these life-forms would even care about us. Since Stephen King's science fiction is a blend of science and horror, he starts with the assumption that aliens would be extremely unfriendly.

As noted, the basic premise of *The Tommyknockers* is that aliens spread disease like nuclear fallout, killing people and turning some of them into replicas of the aliens. In *Dreamcatcher*, aliens infect humans with funguslike byrus, which kills people with Ebola-like certainty. Space invaders killing people with alien diseases is possible. After all, we're only human, and, as such, we're highly susceptible to new infections.

There is no cure for Ebola, an extremely lethal virus that induces massive bleeding in its victims. There's also no cure for Marburg, another extremely lethal virus that causes hemorrhagic fevers. Botulism, a bacteria, can be inhaled or ingested; it causes paralysis and respiratory malfunction. Anthrax is a bacteria that was used widely during the 2001–2002 terrorist attacks on the United States. Smallpox is a virus that is highly contagious and spreads quickly through the air. The list goes on. If unfriendly aliens land on Earth, disposing of the human population wouldn't be very difficult, especially for a race capable of traveling from another world.

Hostile or friendly? Humanoid or not? If aliens from other worlds exist, what does modern science tell us about them? Will they be easy to spot, with motives and ideas just like ours? Or will they be truly alien and be impossible to understand? Looking for answers, we turn to astrobiology, the study of life in space.

The Science of Astrobiology

Let's assume that Carl Sagan's view of the universe is correct and we're not alone in the multiverse of stars. In 2000, the science writer David Darling wrote, "Poised on the brink of a momentous breakthrough that will change forever how humankind thinks about itself and the universe around it, astrobiology is quickly coming of age,"[2] and further, "Almost beyond doubt, life exists elsewhere."[3]

Astrobiology, the study of life in space, is the latest buzzword among scientists who believe that life exists on other planets. Arguments for alien life include the fact that microbes have been discovered on our own planet in areas that we thought could never support life of any kind. There are microbes in volcanic vents in the sea, within the Arctic Cap, and in scorching underground rocks. In addition, scientists have discovered evidence of life on other planets such as Mars. It's now thought by many scientists that life can indeed thrive in alien environments.

In 1998, NASA created the NASA Astrobiology Institute, which clearly implies that very credible scientists believe that alien life exists. The institute's budget was $5 million in 1998, and it rose only to $15 million in 2002. Without more serious backing and space exploration, we may never go forth into space and find aliens, though of course, they may come here and find us—as in Stephen King's books.

As further evidence of blossoming worldwide interest in alien life, astrobiology centers are now in all parts of the world, including Japan, France, England, Australia, and Spain, and two new journals have popped up on the scene: *Astrobiology* and the *International Journal of Astrobiology*.

In 2004, the physicist Michio Kaku told *Astrobiology* magazine:

The question [about whether there are billions of habitable worlds available for evolving complex life] is no longer a matter of idle speculation. Soon, humanity may face an existential shock as the current list of a dozen Jupiter-sized extra-solar planets swells to hundreds of Earth-sized planets, almost identical twins of our celestial homeland. . . . Every few weeks brings news of a new Jupiter-sized extra-solar planet being discovered, the latest being about fifteen light years away orbiting around the star Gliese 876.[4]

Within ten years, we'll be launching the Space Interferometry Mission, which will house many telescopes on a thirty-foot structure. Then, soon afterward, we'll launch the Terrestrial Planet Finder, which will search for other planets as it "scan[s] the brightest 1,000 stars within 50 light years of Earth."[5] The search for alien life continues to fascinate us, despite the fact that we don't really know what we are looking for.

Possible Types of Alien Life-Forms

In 2005, when *National Geographic* asked scientists what aliens might be like, their answers pointed more toward a diversity of species than hostile humanoids. For example, on a planet with extremely high amounts of carbon dioxide, low gravity, and a density atmosphere, trees may take the form of half-mile-high giants with enormous flat tops. On a planet without seasons, days, or nights—that is, a planet orbiting a red dwarf with one side always facing the star—an alien life-form might be an armadillolike creature.[6] As an aside, it's no more likely that alien species would be able to breed with humans than armadillos could breed with humans.

Will we ever find life so strange that we don't even recognize it as life? Might we someday discover a life-form based on silicon, as conjectured in science-fiction stories published in the 1930s, instead of on carbon? Not likely, agree most astrobiologists. The most widely held belief is that alien life-forms will be based on carbon and water as they are on Earth. One of the leading experts on evolutionary biology, Dr. Simon Conway Morris of Cambridge University in England, explained "Life as we know it is extremely special and extremely strange and depends on the remarkable properties of carbon and liquid water."[7]

But if that's the case, there's definitely a possibility that alien life-forms from other worlds, even worlds very different from our own, might resemble us more than we imagine. Maybe the flying saucer people are right after all, and the aliens are hiding just around the corner.

Which leads us to ask, in mainstream science fiction, will the aliens we are going to meet someday be friendly or unfriendly? How better to tell what we expect to encounter in outer space than by judging what we've imagined during the last hundred years?

Aliens We Have Known

The first fictional alien invasion of note occurred in H. G. Wells's 1897 novel *The War of the Worlds*. The invaders are sluglike creatures that use gigantic walking machines equipped with death rays to exterminate as many people as they can. No definite reason is given for the Martian invasion, though it is surmised, as will be the case in many such novels to follow, that they want our planet for its rich biological and mineral resources. The Martians' thinking seems to be based on the idea that having used up everything useful on their planet, instead of dealing with the consequences, it is easier just to invade another world for more. So much for great intellects or brilliant thinkers. The Martians in Wells's novel are quite capable of space travel, building huge destructive machinery, and killing lots of people. They obviously are not into recycling or studying biology, since any fifth-grader could have told them that breathing the air on another planet is the quickest way to cut your own throat when voyaging between worlds.

This lack of practical knowledge seems to be one of the great problems shared by most alien invaders in fiction. Equally annoying, the aliens have a certain smug superiority that leads again and again to their downfall. They are much too eager to explain themselves to any human willing to listen to their evil ramblings. In the 1997 film *Independence Day*, what explanation is there for an alien to take over the body of one of the scientists in the Area 51 laboratory other than being able to brag about how its comrades are going to suck the Earth as dry as an orange? Alien invaders with hubris—it would have made the Greek dramatists proud.

Aliens in Pulp Fiction

It was in pulp fiction magazines published in the United States that alien invaders came into their own. These publications were called "pulps" because they were printed on extremely cheap wood pulp paper. They served as a source of inexpensive fiction for most of the reading public during the period from 1900 to 1955. Over the years, the word *pulp* came to symbolize cheap, worthless fiction, as many of the stories showcased in the pulps were of questionable quality. Still, not everything published there was terrible, and, in fact, many magazines had high standards for their fiction.

One of the top pulps during the first quarter of the century was *The All-Story Magazine*. In 1914, its November 14 through December 5 issues featured *The Empire of the Air*, a serial novel by

George Allan England. In the novel, beings of energy from the fourth dimension, from somewhere beyond the galaxy, invade Earth, hoping to destroy the planet and feed off the energy from the explosion. Fortunately, the hero of the novel has been swallowed by a dimensional rip in space and knows the aliens' plans and how to defeat them. At the end of the story, with most of the energy beings dead, the others leave for their home outside the galaxy. The hero begs one of them to stay, to teach mankind the mysteries of space and time, but, without a word, the weird energy creature disappears.

By the 1920s, alien invasions from outer space had become a staple of pulp fiction magazines. Earth was invaded by living metal aliens that resembled elephants and kangaroos, beings from Venus, more Martians, degenerate cave dwellers from the Moon, intelligent pterodactyls from Venus, batlike humanoids from Mars, microscopic beings from a world inside an atom, giant cockroach creatures from the fifth dimension, and many others. Writers who specialized in Earth-invasion stories included Edmond Hamilton, Ray Cummings, and John W. Campbell.

Campbell produced one of the classics of the genre when he wrote "Who Goes There" (published in *Astounding SF*, July 1938). A group of explorers at the South Pole stumble across an alien spaceship buried beneath tons of ice. The ship is obviously thousands of years old. The one inhabitant on it is not dead, however, but is merely frozen. When the creature thaws out, it escapes. Only after several deaths do the surviving scientists realize that the alien is a metamorph, a creature that can change its appearance to resemble anything—or anyone. The monster has infiltrated their ranks, and unless they can figure out who it is disguised as, it will kill them all. Campbell's heroes use science intelligently to discover the creature's identity. The story was made into a movie twice under the title *The Thing*. In the 1951 version, directed by Howard Hawks, the shape-changing monster is replaced by a blood-drinking humanoid vegetable alien. The film is a masterpiece of suspense, as the small crew is trapped in the Antarctic station while being hunted by a near-invulnerable monster. The 1982 version is much closer to the original story, featuring a shape-changing alien that can duplicate any of the men it kills. The film is a bleak look at what men will do to stay alive.

Aliens in B Movies

By the 1950s, alien-invasion stories had become so commonplace that they primarily served as the basis for grade-B science-fiction films. Movies like *Earth vs. the Flying Saucers*, *Invaders from Mars*, and *Invasions of the Saucer People* provide some thrills and chills but little more. It wasn't until Jack Finney penned the remarkable alien-invasion novel *The Body Snatchers* for Dell paperbacks in 1955 that space invaders regained their sense of menace. Filmed by Don Siegel in 1955 as *Invasion of the Body Snatchers*, the story offers a nightmarish scenario of small-town life in the 1950s, where people are replaced by their exact duplicates, grown from giant pods from outer space. It is a science-fiction and horror masterpiece and is one of the most influential movie thrillers ever made.

Aliens on TV

By the late 1950s, the action had switched from the big screen to the small screen. *The Twilight Zone* took old science-fiction ideas and made them new. Perhaps one of the best alien-invasion stories ever written was penned by Rod Serling as the script for "The Monsters Are Due on Maple Street," which debuted on March 4, 1960.

The setting is Maple Street, somewhere in the United States, around the end of summer. The story begins at dusk. Children are playing outside and adults are talking. The shadow of something unseen passes overhead and a loud roar is heard. Soon afterward, as night falls, the townspeople discover that their power is no longer on and the phones don't work. Even stranger, the radios are dead. The people gather outside to talk things over.

Steve Brand wants to go into the city and see what's happening, but his car won't start. Little Tommy begs him not to go. He's convinced that the power outage is part of an alien invasion, just like he's read about in comic books. Plus, he declares that usually before these invasions begin, aliens disguised as regular people infiltrate the area they want to conquer. Everyone laughs at such an idea, but after time, they start having second thoughts.

Les Goodman tries to start his car but with no success. When he gets out and starts to walk back to the others, his car starts by itself. Suddenly, everyone is suspicious of Les. His neighbors mention how he often stands outside at night looking at the sky. Les complains that he has insomnia. Then the lights in his house start to flicker on and off.

Steve tries to calm the crowd but with no luck. Instead, people grow suspicious about him. As the crowd turns nasty, several people spot a man walking on Maple Street in the dark, coming toward them. One loudmouth, Charlie, grabs a gun and shoots the man. When the people reach him, they realize it was Pete Van Horn, another neighbor who went to see what was happening a few blocks away.

Then the lights go on in Charlie's house. He panics, realizing that now he looks suspicious. Charlie breaks for his house while the rest of the crowd chases him, throwing rocks. Suddenly, all along the street, lights turn on and off and lawn mowers start up and stop for no reason. A riot ensues with everyone shouting accusations at one another.

The view switches to the top of a hill overlooking Maple Street. Two aliens are watching the riot and discussing how easy it is to start such a panic. As one alien says to the other, echoing remarks once made by the comic-strip character Pogo: "They pick the most dangerous enemy they can find and it's themselves."[8]

One can't help but think that Stephen King saw that particular episode of *The Twilight Zone* when he was a teenager and remembered the lesson it taught for the rest of his life. Because despite the horrific monsters in *Dreamcatcher* and *The Tommyknockers*, when it comes to the final analysis, the worst creatures in his books mostly are humans.

Aliens Seen in a New Light

If aliens do land on Earth, it's possible that they will be friendly rather than hostile. For example, in Steven Spielberg's 1982 film *E.T. the Extra-Terrestrial*, ten-year-old Elliott befriends an alien who he names E.T. Stranded on Earth and unable to make it home, E.T. lives with Elliot and his family, and they teach E.T., help him build a device to contact his people on another world, and even enable him to escape government agents. Eventually, Elliot helps E.T. rendevous with his spaceship.

Spielberg's *Close Encounters of the Third Kind* in 1977 also features benign aliens. Although the aliens do abduct humans, they are portrayed as kind rather than evil.

While the aliens do not tend to mate with the humans in space operas such as *Star Trek* and *Farscape*, the aliens are extremely similar to humans: some are kind, some are intelligent, some are friendly, some are nasty, some are less than intelligent, and some are hostile. In these modern science-fiction films and television programs, aliens and humans work together, with the humans being the minority most of the time.

This notion, that intelligent aliens are individuals with good and bad traits, that some will be hostile while others will be friendly, makes the most sense from a logical standpoint. As we discuss more thoroughly later in this book, hostile behavior must be based on something, a root cause. Hostility and evil do not exist in a person at birth. They grow over time due to environmental events, social context, and the person's genetic makeup. For an intelligent, thinking alien, the same would be true. Such a creature presumably is born with a brain containing x number of neurons. A human, for example, has billions of neurons, and each human neuron is connected to thousands of other neurons. In one person's brain, there are trillions of possible neuron states. Why should an intelligent alien lack this type of brain? It makes more sense that like humans, each alien will have a unique brain, one that has millions, billions, or trillions of neurons (or whatever the alien brain cell is called) connected to huge numbers of other neurons, all leading to a unique mind for each alien. With a unique mind, an alien will have—just as a human does—a distinct personality, the ability to decide right from wrong, friendly from hostile, and kind from evil, as the alien sees them. To think that all aliens are evil is to think that all aliens lack intelligence. But clearly, in all forms of film and media, aliens are intelligent, for they create spaceships and reach Earth long before mankind reaches them.

In fact, we can only hope that the aliens are more intelligent and friendly than mankind. Humans tend to be territorial and aggressive. We tend to fear what we don't know. We are wary of outsiders.

America's First Alien Invasion

North and South American Indians probably viewed the Europeans as aliens, but rather than fear Europeans, the Indians embraced them. In the long run, it would have been better had Indians viewed the alien Europeans as invaders.

In the sixteenth and seventeenth centuries, when Europeans began arriving in the Americas, the Indians welcomed them as friends. Indians were fascinated and delighted by the beards, strange clothing, ships, steel knives, swords, cannons, mirrors, and brass kettles.

Sadly, Europeans considered Indians a commodity, just as they viewed the forests, the beavers, and the buffalo. The newcomers tried very hard to convert the Indians to Puritan and Jesuit ways. Of course, the Indians were repelled by the arrogance of their visitors and worried about the exploitation and the destruction of nature. Indians began to see Europeans as hostile invaders.

The conflicts led to the Indian Wars, the Indian Removal Act, and eventually to the massacre at Wounded Knee, South Dakota. The Indian Wars encompassed approximately forty wars between American Indians and the white colonialists from 1775 until 1890. President Andrew Jackson signed the Indian Removal Act into law in 1830. For ten years, seventy thousand Indians had to leave their homes, give up their land, and move to designated places in Oklahoma. They followed trails that became known collectively as the Trail of Tears, and countless Indians died on those trails. In 1890, five hundred government troops tried to round up an encampment of Lakota Sioux Indians in Wounded Knee, South Dakota, and transport them to Omaha, Nebraska. A skirmish resulted in an outright massacre, with 153 Indians dead, 62 of them women and small children.

Let's hope that when the space aliens come to Earth, they treat us better than Indians were treated by the Europeans, who actually believed they were serving their fellow man. And let's hope that we're as trusting, if perhaps more circumspect than the Indians were. Just read any Stephen King novel to witness humanity's distrust and hatred of the outsider. Maybe that's why King's books are so popular. They play on our innermost fears and our collective human terror of the unknown, including strangers.

4

THE FOURTH HORSEMAN

The Stand

He was scared, so deeply scared he hardly dared admit to
himself. It was the kind of fear that could drive you mad.

—*The Stand*

Humans have battled plagues through history. In the 1300s,
we had the Black Death. In the early 1900s, the Spanish had a
deadly form of avain influenza. And in 1978, we had Stephen
King's superflu.

Coughing to the Apocalypse

Stephen King's concept of a worldwide plague wiping out most of
the human race was not new with his 1978 novel *The Stand*. The
earliest known story about the subject was Mary Shelley's *The Last
Man*, first published in 1826 to mostly terrible reviews. Critics
found the novel too depressing and "godless." It was banned in

Austria and dismissed by most intellectuals of the time as being far inferior to *Frankenstein*. Taking place in 2073, *The Last Man* describes a virulent plague that wipes out all of humanity except for the book's narrator, who at the end of the story sails off into the sunset.

It was H. G. Wells who first recognized that biology could serve as a deadly weapon. In his novel of Earth's invasion from outer space, *The War of the Worlds* (Heinemann, 1898), the near-invincible sluglike Martians aren't defeated by human weapons but instead perish to common disease bacteria in the atmosphere.

Perhaps the most effective description of a biological weapon gone berserk published up until *The Stand* is Wilson Tucker's grim novel *The Long Loud Silence* (1952). In the novel, the United States is sneak-attacked by a nameless foreign power using atomic and biological weapons. The few survivors, including the hero, find themselves isolated on the east side of the Mississippi River, with soldiers on the opposite bank refusing to let them cross for fear that they could spread plague germs. The brutal tale of civilization sinking back into barbarism and cannibalism is presented in an understated but grimly effective manner.

Richard Matheson, who served as one of King's strongest literary influences, composed a frightening novel of biological warfare in the near future in *I Am Legend* (1954). In the story, a deadly artificial plague infects mankind, and there is no cure. Only a few humans are immune to the disease that turns the rest of the population into vampires who feed on blood. Matheson's description of the breakdown of civilization was amplified by King to great effect in *The Stand* a quarter of a century later.

In the 1960s and 1970s, the atomic bomb was the weapon of choice in novels of humanity's destruction. Biological weapons seemed very hit or miss when compared with the total devastation brought about by an atomic bomb. Plus, zombies created by atomic radiation were all the rage. It wasn't until the publication of *The Stand* that plague novels returned to their earlier popularity. Expanding on the end-of-the-world scenarios of *The Long Loud*

Silence and *I Am Legend* and adding an unexpected supernatural twist, King wrote what many fans and critics considered his best book. It was definitely one of his most frightening.

The unabridged edition of *The Stand* begins with a short prologue, "The Circle Opens." Charlie Campion, a security guard at a top-secret government research center in California, witnesses an accident in the plague research lab that releases a deadly super-flu virus throughout the base. Due to a malfunction in the security net, Charlie escapes from the base. Gathering up his wife and child, he drives east, unknowingly carrying the plague virus in his body.

The novel proper begins a day later. Base security fails to notice Charlie is gone until nearly twenty-four hours have passed. During that time, he's made it to the little town of Arnette, Texas. His wife and child are already dead in the back seat of the car. Several locals, including Stu Redman, try to talk to Charlie before he, too, dies, but they can't make much sense out of what he says. It hardly matters. Everyone at the gas station is exposed to the superflu. Worse, the cousin of the gas station owner is state policeman Joe Bob Brentwood. When he comes around the next day to warn his cousin that investigators from the Plague Center in Atlanta, Georgia, have flown into town to investigate the deaths, he's also exposed to the virus.

Later that day, Joe Bob pulls over a speeder named Harry Trent, an insurance salesman, and gives him a ticket. Without realizing it, Joe Bob passes on the plague to Harry. What happens next defines the rest of Book 1.

Harry, a gregarious man who likes his job, passes the sickness to more than forty people during that day and the next. How many those forty pass it to is impossible to say; you might as well ask how many angels can dance on the head of a pin. If you were to make a conservative estimate of five apiece, you'd have two hundred. Using the same conservative formula, one could say those two hundred went on to infect a thousand, the thousand to infect five thousand, and the five thousand to spread it to twenty-five thousand. Under the California desert and subsidized by the taxpayer's money, someone had finally invented a chain letter that really worked—a very lethal chain letter.[1]

The superflu virus, formally known as Project Blue by the military men who created it, is dubbed by the trendy in-crowd of California as Captain Trips. Whatever name it is given, the flu virus proves to be 99.4 percent effective. Captain Trips, for people who aren't historians of the Age of Aquarius, comes from the nickname of Al Hubbard, one of the early promoters of LSD use. It was also the nickname of the Grateful Dead lead singer Jerry Garcia.

Hardly anyone who is infected with the superflu recovers. During the course of the next nineteen days, the plague wipes out just about everyone in North America. When it becomes obvious to the commanders of the army that the United States will be helpless against foreign invaders, a top military man instructs overseas agents to release the plague germs among the unsuspecting populations of Europe and Asia. It's never actually stated in the book, but it's assumed by readers that the entire world is equally decimated. It's worth noting that even if the plague was 99.4 percent lethal, in 1978's world of slightly more than 4.3 billion people, this would still mean that nearly 260,000 people would survive the plague. In the United States, with a 1978 population of more than 222 million people, a .006 survival rate would suggest that just well over 12,000 people would survive. That's not much, but it's a lot more people than the number who seem to be around in the second half of *The Stand*. Which points out the danger of mentioning an actual statistic when telling a horror story.

The second and third parts of *The Stand* describe how two very disparate groups of survivors make their way across the country to two very different locations. It is the conflict between these two remnants of civilization, good and evil, that defines the rest of the story, as mankind faces its final stand.

The novel ends with Stu asking Frannie, another survivor of the conflict, whether the people who have survived have learned anything, if people can ever learn from their mistakes. Fran answers, "I don't know."[2]

Start Spreading the News

The Stand is Stephen King's most gripping and horrifying book because the first part of the novel reads more like nonfiction than fiction. The spread of the Captain Trips is relentless and inevitable. People get the virus, pass it on to other people, then die. Unlike *The Long Loud Silence*, there is no river to stop the disease's spread. If anything, the most unbelievable scene in the novel is when several military men discuss whether they should infect Europe and Asia with the plague. The implication of the conversation is that the people of those continents haven't been infected by the disease. With today's constant stream of people traveling between one continent and another, it seems unlikely that anyone anywhere would be safe from the superflu. We live in—as we are reminded constantly by airlines and phone companies, newspapers and politicians, banks and teachers—a global community.

Is what happens in *The Stand* possible? Or even worse, is it probable? How realistic are the events in the first part of the book? Let's examine the spread of the disease and see exactly what happens and how fast it takes place. Does the government act intelligently or foolishly to prevent the end of the world?

Captain Trips, the superflu developed by the scientists of Project Blue, is nearly 100 percent effective. It has an incubation period of one day and then attacks and kills the victim within a few hours. Most diseases don't remain lethal after being transmitted from one donor to the next three or four times. Their effect diminishes with the age of the virus. Yet because Captain Trips constantly mutates from one flu strain to another, it remains deadly long after most other plague germs have lost their potency.

In the novel, Charlie Campion flees the secret base where Project Blue is being developed, taking his family with him. By the time Charlie arrives by car in Arnette, Texas, the next day, his family is dead and he is near death. He survives just long enough to infect the men at the gas station.

Now, if the government had really been on the job, the plague would have stopped then. Having Charlie escape the dying base seems somewhat unlikely but possible. Not noticing that Charlie wasn't among the dead until twenty-four hours later is highly improbable. Anyone investigating the deaths at the lab would know that containing the plague is the most important job, so a comprehensive tally of the bodies on the base would have been the number-one priority. Charlie and his family should have been discovered long before they arrived at Arnette. But that doesn't happen in the novel, and the plague is thus allowed to spread.

The government health agency wisely quarantines the entire population of Arnette. But again, there's a major mistake of omission in that no effort is made to locate any people who were in Arnette shortly after the plague arrived and who moved on. Thus, the government allows the plague to leave the town and infect the general population. Isolating the plague and potential carriers is the most important task in preventing a health disaster, and in that respect, the government in *The Stand* fails dismally. Of course, *The Stand* was written in the 1970s, and since then we've tightened up security throughout the United States in regard to biological weapons—at least, since September 11, 2001. As expressed by William C. Patrick III during a Washington roundtable on science and public policy:

> The sample case I brought with me today holds glass bottles containing exact simulants of the weaponized form of anthrax and the virus causing Venezuelan equine encephalitis virus. Now I've carried this case through a number of airports on my way to meetings over the last twelve years, but no one ever stopped me and asked "What are those peculiar looking powders?"
>
> Then I started carrying crude disseminators, like this small plastic spray bottle. It's just a single-fluid nozzle and it doesn't generate a great deal of pressure; the liquid it sprays produces large particles that fall out of the air very quickly. But a liquid agent is very easy to make. It's something that a terrorist might attempt to use. Yet, when I travel through airports, no one has ever stopped me.[3]

Such would not be the case today. Or at least we hope not.

Charlie infects the service station owner Bill Hapscomb, who infects his cousin, the policeman Joe Bob Brentwood, who infects the traveling salesman Harry Trent, and what is known as a geometric progression begins. The power of a geometric progression is what makes *The Stand* so frightening.

An arithmetic progression is a sequence of numbers where the difference of any two successive members of the sequence is a constant. Thus, the sequence 1, 2, 3, 4 . . . is an arithmetic progression where the difference between any two successive members is 1. The sequence 2, 5, 8, 11, 14 . . . is an arithmetic progression with the difference being 3. Arithmetic sequences can be thought of as being built by addition.

A geometric progression is a sequence of numbers where the quotient of any two successive members in the sequence is a constant called the common ratio of the sequence. A geometric progression is written as ar^0, ar^1, ar^2, ar^3, ar^4, ar^5, . . . where a is a number called the scale factor, and r is the common ratio. In a geometric progression, r cannot equal zero. In math, $r^0 = 1$, so $ar^0 = a$; $r^1 = r$, so $ar^1 = ar$. A typical geometric progression is 1, 2, 4, 8, 16, 32 . . . where $a = 1$, $r = 2$. Another geometric progression is 2, 10, 50, 250, 1,250, . . . where $a = 2$ and $r = 5$. A geometric sequence can be thought of as being built by multiplication by r.

Now, real life isn't governed by geometric progressions, but we can simplify real life by assuming that human interactions follow simple mathematical models. It's only a short step from geometric progressions to the notion of "six degrees of separation."

The concept of people being connected by social links was first laid out in a 1929 short story named "Chains" by the Hungarian writer Karinthy Frigyes.[4] It later became known as the small-world effect. The idea extends geometric progressions to their logical conclusions. The phrase "six degrees of separation," comes from a 1967 experiment by the psychologist Stanley Milgram, who hypothesized that any two people in the United States could be connected by a chain of six friends.

"Six degrees of separation" is best illustrated by the game Six

Degrees of Kevin Bacon. In the game, contestants try to link a particular actor to the actor Kevin Bacon through the use of connecting movies as steps. For example, a player might be asked to link Dwayne "the Rock" Johnson to Kevin Bacon in the least number of steps. One possible solution would be that The Rock starred in *The Scorpion King*. Michael Clarke Duncan was also in that movie. Duncan was in *The Green Mile*. Tom Hanks also starred in *The Green Mile*. And Tom Hanks starred with Kevin Bacon in *Apollo 13*. Therefore, there are three degrees of separation between The Rock and Kevin Bacon.

Let's construct a few geometric progressions using Harry Trent as the first person in our chain. According to *The Stand*, Harry passes the superflu on to forty people. In our geometric progression, $a = 1$ and r would be 40, so $ar^0 = 1$ person and $ar^1 = (40)^1 = 40$. Now, if life imitates a geometric progression, each of the 40 people Harry meets will infect 40 more in the next day or so, resulting in $ar^2 = (40)^2 = 1{,}600$ people. If the geometric progression continues (with $a = 1$), then $r^3 = (40)^3 = 64{,}000$ people. Now, according to *The Stand*, it takes nineteen days for the superflu to wipe out the population of the United States. If we assume that each person infected with the flu meets forty people the day after he or she is contaminated, and if Harry Trent begins our geometric progression on the first day of the flu, then it follows from our formula that on day nineteen, the total number of people infected will be $(40)^{19} = 2{,}748{,}779{,}069{,}000{,}000{,}000{,}000{,}000{,}000{,}000{,}000{,}000{,}000$, a number vastly bigger than the known atoms in the universe. Since there are nowhere near that many people on Earth, we are forced to conclude that everyone on earth will have been exposed to the plague millions of times. The only exceptions are hermits who have had no contact with any human being for years at a time. Otherwise, any person with the least amount of contact with other people will be infected. It will be the end of life on Earth as we know it.

But not every person knows or meets forty people in a day. So, for the sake of argument, let's say that Harry Trent merely meets and infects five people instead of forty. And, after that, those five people meet only five people each. So our geometric progression turns into 1, 5, 25, 125, where $a = 1$ and $r = 5$. What will be the result when we reach the nineteenth member of the geometric chain?

It will be another casse of overkill: $(5)^{19} = 19,073,486,328,125$ or, in approximate numbers, in nineteen days the plague will have spread to nineteen trillion people. Again, more than enough exposures so that everyone on Earth will have been infected 1,000 times over with the superflu.

Once a geometric progression starts in earnest, there's nothing in the world that can stop it, other than nipping the sequence off at one of the early stages. In other words, the plague has to be contained immediately, or it cannot be stopped. Math doesn't lie.

If there's a secret laboratory somewhere in the Rocky Mountains and government scientists are working on a deadly strain of flu virus for which no antidote exists, and a window breaks and someone infected with the disease runs away, then the world is doomed. Unbelievable as it sounds, it could happen. On a somewhat smaller scale, with a slightly less virulent virus, it already has—more than once.

The Stand in History

The greatest disaster in recorded history took place during the years 1347 through 1350. It was a disease outbreak that became known as the Black Death, and it killed an estimated 34 million people, approximately one-third of Europe's population. Records from the Far East and the Middle East show that the Black Death was part of an even larger bubonic plague pandemic (a pandemic is defined as an epidemic over a wide geographic area and affecting a large percentage of the population) that struck much of Europe, Asia, and Africa. The total number of people killed by this pandemic will never be known, but some historians estimate that, in total, more than 60 million people died due to the plague.[5]

Bubonic plague, the main cause of the Black Death, returned to haunt Europe again and again until the beginning of the eighteenth century. Recurrent episodes of the Black Death included the Italian

Plague of 1629–1631, the Great Plague of London (1665–1666), and the Great Plague of Vienna (1679).

The only plague comparable to the Black Death was an unnamed plague that began in Egypt in A.D. 451. This plague swept across the ancient world over the next four years and reportedly killed half the population of the countries it struck. The exact number of deaths from that earlier disaster probably was never known.[6]

According to modern researchers, the Black Death most likely began in the steppes of central Asia, although some historians believe it might have originated in northern India. The cause of the disease was a bacteria named *Yersinia pestis*, which was carried and spread by fleas. Plague fleas were transported by rats across half the world. It was the unchecked spread of rats through Asia, Europe, and the Middle East that brought the Black Death. No other bacteria had so much of an effect on human history.

There were three types of plague. Bubonic plague was the most common. A flea bite deposited the bacteria into the victim's lymphatic system. The disease was characterized by buboes, large, inflamed, and painful swellings in the lymph glands of the groin, the armpits, or the neck, depending on where the flea bite occurred.

In septicaemic plague, which was almost always fatal, the bacteria entered the bloodstream directly, rather than through the lymphatic system where they might be contained. Like bubonic plague, the septicaemic variety of plague was caused directly by flea bites. Death usually took place within twenty-four hours of catching the disease.

Pneumonic plague was the most deadly form of plague. It was usually fatal and wasn't cause by a flea bite. When the plague bacteria reached the lungs of a victim, this caused severe pneumonia. The bacteria were present in water drops spread by coughs and choking. This third variation of the plague was highly contagious. Death from the pneumonic plague occurred within three or four days.

In all three versions of the plague, internal bleeding caused large bruises to appear on the skin. People infected with the plague suffered from acral necrosis, a symptom of the disease where the victim's skin turned black due to subdermal hemorrhages. This bruising resulted in the plague being called the Black Death.

Whether the plague originated in northern India or on the steppes of Mongolia, it was brought east and west by traders and soldiers along the trade route that linked Europe and China known as the Silk Road.

Conditions in both Europe and China were exactly right for the spread of bubonic plague. The poor were extremely poor, sanitation was unheard of, and bathing rarely ever took place. Needless to say, rats and fleas were everywhere. Nor did the few scientists of the time have any idea how diseases were spread. The notion that the plague could be spread from one location to another by carriers was never even considered.

A civil war had raged in China between the native population and Mongol invaders since 1205. The war disrupted normal planting and farming and led to widespread famines among the population. Meanwhile, in Europe, during the years 1315 to 1322, a major famine enveloped the Continent and food shortages were commonplace. Grains were in short supply, and thus livestock were usually thin and malnourished. Hungry workers did less work, resulting in less grain, which brought about even higher prices, thus making it even more difficult for workers to buy food.

Along with suffering from poor nutrition and famine, the people of Europe and Asia lived through some of the worst weather in recorded history during the first half of the fourteenth century. Massive rainstorms pelted the lands year after year, and summer barely took place. Long, very cold winters sapped what little strength and vitality the poor possessed. Thus, most people of the time were already in bad health and vulnerable to disease when the Black Death hit. They made easy targets for the virulent bacteria.

In 1318, an unknown disease, now thought to have been anthrax, devastated the animal population of Europe. The disease killed sheep and cattle, making meat scarce and cutting into the income of the peasant class. The poor grew even poorer and more unhealthy.

Meanwhile, in Asia, the bubonic plague struck China in 1334.

The plague traveled with merchant caravans to the West over the next decade. In 1347, bubonic plague was reported in the trading centers of Constantinople and Trebizond in 1347. That same year, the city of Caffa, a seaport on the Crimean peninsula, was attacked by the Tatar horde of Janiberg. The Italian city-state of Genoa supported Caffa, while its trade rival, the city-state of Venice, assisted the Mongols. The horde put the city to siege, but soon afterward, plague struck the attackers and killed thousands. So many men died that bodies were stacked like firewood against the walls of the city.[7] The Tatars broke off the siege, but the disease had already spread to the city.

In October 1347, a fleet of Genovese ships escaping from Caffa docked at the port of Messina. All of the crew members of the ships were either dead or dying from the plague. Most likely, the ships carried infected rats and fleas, which spread the disease to the city. Looters unwittingly helped to bring the plague into town. Messina was soon crippled by the plague, and people fleeing the city quickly passed the disease on to Genoa and Venice. For the curious, a great fantasy adventure novel, *Red Eve* by H. Rider Haggard (1912), describes in vivid detail the mysterious arrival of the plague in Genoa, Italy.

Moving from Italy, the plague carved a swath of death across Europe, striking France, Spain, Portugal, and Great Britain by June 1348. It then spread eastward through Germany and Scandinavia in the years 1348 to 1350. In 1351, it struck northwest Russia. Strangely enough, the plague hardly touched Poland. Modern scientists think that at the time, the Polish population was so scattered and isolated that the plague couldn't spread from person to person as it did in most European countries.

The pandemic hit the Middle East in 1347. The muslim leader Malik Asraf returned with his troops from a battle near the city of Tabriz in Russia, where the plague was raging. Asraf directed his army to attack the town of Hasan Buzurg, near Baghdad, but plague struck the soldiers and decimated their ranks. Soon afterward, the plague, no doubt carried by some of the troops, hit Baghdad.

In autumn 1347, the Black Death struck Alexandria, the disease most likely spread by rats that jumped off trading ships from Constantinople. The next year saw the plague spread across most of the major cities of the Middle East. In the winter of 1348–1349, the disease attacked Antioch. Many of the city's inhabitants fled to the north, thus spreading the Black Death into Asia. By then, its effects in Europe were coming to an end, and the great burial and renewal of life began.

Still, the plague repeatedly returned to haunt Europe and the Mediterranean throughout the fourteenth to seventeenth centuries. The Great Plague of London in 1665–1666 was generally recognized as one of the last major outbreaks. Many speculate that the Great Fire of London in 1666 may have killed off any remaining plague rats and fleas, which led to a decline in the plague. The destruction of rats in the Great Fire may also have contributed to the ascendancy of brown rats in England. The bigger Norwegian, or brown, rat was not as likely to carry germ-bearing fleas, which were the cause of the earlier plagues.

The Black Death changed the course of history. With so many people dead, there weren't enough peasants to work the land and harvest the crops. Thus, landowners had to offer better wages and living conditions to convince serfs to work for them. This early competition for workers signaled the end of the feudal system. In a sense, the Black Death brought about the start of capitalism.

The church, a strong player in keeping the peasants in line, was immobilized, with so many of its priests and nuns killed by the plague. Plus, peasants and nobles alike saw firsthand that monks and priests were unable to save anyone, including themselves, from the Black Death, so their faith in the church was severely shaken. The stranglehold the church had around the necks of European peasants was broken, never to return.

At the end of *The Stand*, Stephen King makes it clear that the plague that wiped most of the earth clean of life had also given mankind a new chance, a second chance. What happens next is left to the reader's imagination.

It's difficult to comprehend in these modern times how the Black Death spread so swiftly from city to city. That's because we think people will act logically and worry more about the group than about themselves. That's definitely not the case in real life. *The Stand* offers a striking example of how a plague travels. The big difference between *The Stand* and the Black Death is that the people of Europe didn't know how to act. In our modern world, we know what we should do, but the big question is, will we do it? In *The Stand*, the action of one man, Charlie Campion, condemns nearly the entire human race to death. It's a very frightening thought because we know it's true. When it comes to chemical or biological warfare, the actions of one individual can make all the difference in the world.

The Superflu in the Twentieth Century

King's scientists are working on a "super" flu, but he might just as well have used the common flu, which kills around 36,000 Americans every year. And less than a hundred years ago, a flu outbreak killed more than 50 million people. It wasn't *The Stand*, but for many people of the time, it must have seemed awfully close.

The Spanish flu pandemic was an extremely deadly form of avian influenza that killed somewhere between 50 and 100 million people over a period of approximately twelve months during 1918 and 1919. The flu was caused by the H1N1 type of influenza virus. H1N1 is a subtype of the species called avian influenza virus (bird flu). Avian flu is a disease, and avian flu virus is a species. The avian flu virus subtypes are labeled according to an H number and an N number. The H1N1 virus is similar to the bird flu of the present, which is labeled H5N1 and H5N2.

During World War I, the Allied Forces called the H1N1 flu the "Spanish flu" because Spain was not involved in the war and thus there was no wartime censorship. The flu generated a lot more press in Spain than in any country involved in the fighting. There was no evidence that the flu came from Spain, but more than 8 million people were infected with it in May 1918. As in *The Stand*, public officials in Spain assured the population that the flu was nothing dangerous.

As mentioned previously, the most likely cause of the pandemic was an H1 avian virus. The cause was not determined at the time because the influenza virus was not understood by science of that period.

A popular theory suggests that the virus type originated at Fort Riley, Kansas, due to the chickens and pigs that were bred at the fort for food. Poultry and swine are normally affected by different strains of influenza, but they can be infected by a strain not typical of the species if a random mutation takes place in one of the strains. Scientists have theorized that a strain of flu that was jumping between pigs and chickens may also have infected humans. When such a shift takes place, the mutated flu becomes a new species of human flu, one that the body has no built-in antibodies to combat. Such flu mutations are incredibly contagious.

Recent studies suggest that the flu at Fort Riley may have jumped directly from chickens to humans, most likely to the cook at the fort. Whether the flu came directly from birds or from pigs hardly matters. In either case, the flu strain proved deadly.

The first case of flu at Fort Riley was reported on March 11, when a soldier showed up at the camp hospital complaining of fever, sore throat, and a headache. He was the first of many to become ill that day. By noon, the hospital had seen more than a hundred sick troops. Within a week, that number stood at five hundred. It took months for the flu to spread throughout the country, but by September, it had reached epidemic proportions.

As with the Black Death, the general health of the soldiers fighting in World War I might have played a major role in the spread of the flu. It's been suggested that troops were more vulnerable to the flu due to the stress of trench warfare, along with the dangers of chemical weapons. Plus, the close quarters of living in the trenches and the movement of troops from one area to another surely helped to spread the disease faster than would have happened in normal living spaces.

The flu was deadly and killed without mercy. The strain was unusual in that unlike normal influenzas that killed primarily the old and the very young, it attacked the young and the healthy. People without symptoms suddenly got sick, within a few hours became too worn out to walk, and died the next day. Symptoms included coughing blood and a blue tint to the face due to obstruction of the lungs. Later on, the virus caused an uncontrollable hemorrhaging that filled the lungs. Patients literally drowned in their own body fluids.

In September 1918, the Spanish flu turned into a worldwide pandemic and during its first twenty-five weeks killed an estimated 25 million people. Some historians believe that the flu might have been one of the factors that led to the end of World War I. It's been estimated that at least 20 million people (or more) on each side of the war died from the flu during the last months of the battle. More people died from the Spanish flu during World War I than were killed in the fighting. In his novel *The Long Loud Silence* (1952), author Wilson Tucker effectively describes the United States of the future in the throes of a major pandemic, modeled in many ways after the Spanish flu.

Modern researchers put the global mortality rate from the 1918 flu epidemic at somewhere between 2.5 and 5 percent of the human race, with nearly 20 percent of the world's population suffering from the disease to some measure. Some social scientists consider these figures extremely conservative and believe that the flu may have killed nearly 100 million people. In the United States, about 28 percent of the population suffered from the flu, and more than 500,000 died. Almost 200,000 deaths were recorded in the month of October 1918 alone. In Great Britain, 200,000 people died, while in France, the number of deaths was more than 400,000. The death rate was extremely high for native populations with little immunity to major diseases. Entire villages of Aleutian islanders in Alaska were wiped out. Among the Fiji Islands, 14 percent of the population died in two weeks. In Western Samoa, the death rate was more than 20 percent.

Needless to say, the social effects of the flu were felt all around the world. While in most places less than one-third of the population was infected and a fraction of that died, in a number of towns in several countries the entire populations were wiped out. Many cities, states, and countries enforced restrictions on public gatherings and travel to try to stop the pandemic. In some towns and cities, theaters, dance halls, churches, and other public gathering places were closed for longer than a year. On October 2, 1918, Boston registered 202 deaths from the flu. Soon afterward, the city canceled its Liberty Bond parades and sporting events. Churches were closed, and the stock market was put on half-days. In New York, 851 people died of influenza in a single day. In Philadelphia, the city's death rate for one week was 700 times higher than normal. The crime rate in Chicago dropped by 43 percent, with police attributing the drop to the number of criminals who had the flu.

Quarantines were enforced with little success. In November, at the end of World War I, thirty thousand San Franciscans went outside to celebrate. There was much dancing and singing. Everybody wore a face mask.[8] Some communities placed armed guards at the borders and turned back or quarantined any travelers. One U.S. town even outlawed shaking hands.[9]

Even in areas where mortality was low, those incapacitated by the illness were often so numerous as to bring much of everyday life to a stop. Some communities closed all stores or required customers not to enter stores but to place their orders outside the stores for filling. There were many reports of places with no health-care workers to tend the sick because of their own ill health and no able-bodied grave diggers to inter the dead. In many places, mass graves were dug by steam shovels and bodies buried without coffins.

The Spanish flu, having run its course around the world, vanished within eighteen months.

The Superflu Today: The Threat of Bird Flu

The Spanish influenza pandemic took place approximately ninety years ago and killed millions of people. Since then, the world has suffered through two more, much milder, flu pandemics. In 1957–1958, the "Asian flu" was the second pandemic of the twentieth century. It was caused by an H2N2 virus. The Asian flu started in China and killed a million people around the world, including seventy thousand in the United States.

In 1968–1969, the "Hong Kong flu" caused the third pandemic of the twentieth century. It was caused by an H3N2 virus and killed some thirty-four thousand Americans in one winter.

From the time of the Spanish flu to the present, scientists have worked hard on trying to isolate, understand, and cure the avian flu. The bird flu has proved to be an extremely hard disease to eradicate. Researchers know that unless they find a cure for the disease, it will strike again. And while our health services are better, our transportation systems have improved even more. Stopping the spread of the disease will be nearly impossible unless the government quarantines entire cities—which, most likely, wouldn't work.

As far back as the late 1920s, Richard Shope demonstrated that swine flu could be transmitted through filtered mucous, which implied that flu was caused by a virus. In 1933, Sir Christopher Andrews, Wilson Smith, and Sir Patrick Laidlaw were able isolate the first human influenza virus. Then, in 1940, Frank Macfarlane Burnet was able to grow the flu virus in a laboratory. It wasn't until the mid-1970s, however, that researchers realized that enormous amounts of influenza virus continuously circulated in wild birds.

An outbreak of HPAI (Highly Pathogenic Avian Influenza) took place in 1983. The avian flu was caused by the H5N2 virus, but it didn't spread among humans; however, the disease attacked chickens, turkeys, and guinea fowl in Pennsylvania and Virginia. The virus was finally brought under control after the destruction of 17 million birds.

In 1996, HPAI H5N1 bird flu was found in a goose in Guangdong, China. A year later, in May 1997, the virus struck the poultry markets in Hong Kong, causing an epidemic. The first human death was reported from H5N1 soon afterward. By November, there were eighteen more cases of people with the bird flu. Six of them died. Officials had 1.4 million chickens and ducks killed.

Approximately five years later, on January 5, 2003, eleven healthy children in Vietnam suddenly needed hospitalization due to severe respiratory illness. At first, no one seemed sure what caused them to become sick. Then, on January 8, 2003, government offi-cials in Vietnam announced the outbreak of bird flu H5N1 at farms in southern provinces. Around seventy thousand birds were destroyed. Within a month, seven of the children affected died. Tests performed on two of the dead children confirmed that they were infected with H5N1 avian flu.[10]

By the end of January, cases of H5N1 avian flu were reported in Japan, Thailand, Laos, Pakistan, Cambodia, and China. The World Health Organization (WHO) labs determined that the current flu virus was different from the one in Hong Kong in 1997. Since that time, the flu had mutated into a different virus. Millions of chick-ens were killed to try to stop the spread of the disease.

In February 2003, approximately 36 million birds in southeast Asia were destroyed, either killed by the flu or killed to prevent the spread of the virus. Meanwhile, on February 8, U.S. officials reported a case of H7 avian flu at a farm in Delaware. As a precau-tion, twelve thousand birds were destroyed. Another outbreak was reported at a second farm in Delaware, and seventy-two thousand birds were destroyed. Within days, Japan, China, Poland, Malaysia, Singapore, and the Republic of Korea banned poultry imports from the United States.

By April 2003, the virus seemed to have run out in humans. During the initial outbreak of the disease, less than a hundred people died. Fortunately, the avian flu never turned communicable between people. Everyone who got the disease contracted it from birds.

Then, in December 2003, tigers and leopards in a zoo in Thailand died from H5N1 bird flu after eating fresh chickens. It was the first time ever that big cats were known to contract the flu.

The same month, authorities in South Korea reported the pres-ence of bird flu among the country's birds. No humans were infected with the disease, but 1.3 million chickens and ducks were destroyed.[11]

In 2004, H5N1 appeared again in Vietnam, China, and Thailand, as well as in Russia, Turkey, and Romania. Millions of birds were killed, with only a few humans dying.

The last outbreak of bird flu in the United States occurred in February 2004. A flock of chickens in Texas came down with the H5N2 virus. Fortunately, quick action by state and federal officials kept the disease from spreading. No human cases of the flu were reported. Still, no one felt safe as long as there was no antidote to this dangerous strain of the disease. More than half of the cases involving infected humans resulted in death.

The year 2005 proved to be little relief, with more cases reported across the world and no vaccine working effectively. In November, when a duck from a poultry farm in British Columbia was discovered to be carrying the H5 virus, the United States put an interim ban on poultry exports from that Canadian province.

On November 30, 2005, the WHO and a number of regulatory bodies from various countries, including the U.S. Food and Drug Administration (FDA), announced plans to meet in early 2006 to discuss how to speed up production of a bird flu vaccine.[12] Around the same time, international health experts warned that the official numbers of bird flu deaths may be too low and governments may be greatly underestimating the problem.

Meanwhile, during the first few months of 2006, a flood of phony bird flu cures appeared for sale on the Internet. So far, the only cure for the disease is to isolate the population that has the flu and let it burn out without outside interference. Because of the flu's high rate of mutation, it's been estimated that any vaccine developed for bird flu would not be available for at least six months after a pandemic started.[13] A high-level cabinet memo that circulated among members of the British government estimated that if the bird flu struck Britain, seven hundred thousand people would die before the government could take any measures against the disease.[14] *The Stand* may be closer than anyone imagined.

The Real Project Blue

Biological warfare, also known as germ warfare, is the use of any organism, including bacteria, a virus, or some other disease-causing organism or poison found in nature, to wage war. Biological warfare is designed to kill enemy soldiers and, in some cases, enemy civil-ians. Biological warfare also means attacking nature in the area where the enemy is located, destroying his food supplies, destroy-ing his environment, destroying his habitat.

The creation and stockpiling of biological weapons was out-lawed by the Biological Weapons Convention of 1972, which was signed by more than a hundred countries. Biological weapons were deemed too extreme for warfare, as they could cause deaths in the millions and major economic disasters in countries throughout the world. In a strange bit of wording, the treaty prohibited the creation and the storage of the weapons but did not outlaw the use of these weapons. Still, most military analysts consider biological warfare not to be a good mode of attack; it is useful only as a terrorist tool.

Biological warfare is considered useless by most armies. It nor-mally takes days for the full effects of a disease to devastate a city, whereas a nuclear attack would take minutes or hours at most. A biological attack would not stop an advancing army if the men in the army were prepared for chemical weapons. Using poison in the air is dangerous, as there is no way to control air currents, and what would be deadly for inhabitants of a city would poison the invading soldiers as well. Plus, the use of biological weapons usually incurs retaliation in the same form, making it impossible for both armies to breathe. Despite all of these reasons for not resorting to biolog-ical weapons, they have been used throughout history—and most likely will be used again in the future.

Biological warfare existed as far back as the sixth century B.C., when the Assyrian armies poisoned enemy wells with a mind-altering fungus that would drive their enemies mad. In 184 B.C., Hannibal of Carthage had his army fill clay pots with poisonous snakes and instructed the soldiers to throw the pots onto the decks of Pergamene ships.

During the Middle Ages, the Mongols threw diseased animal bodies into the wells and the rivers used by their European enemies for drinking water. Before the Black Death hit all of Europe, Mongols were notorious for catapulting diseased corpses into cities they besieged, hoping to infect the population with the plague.

The practice of throwing the corpses over city walls only grew worse after the plague enveloped Europe. The last time infected corpses were used as weapons of terror was at the beginning of the eighteenth century.

Of course, white settlers often gave Native Americans blankets and clothing that had been exposed to smallpox, a disease that the natives had absolutely no immunity to. At least once, the British used smallpox to infect the Lenape Indians by giving them infected blankets during Pontiac's War.

The use of biological weapons was banned in international law by the Geneva Protocol of 1925. The 1972 Biological and Toxin Weapons Convention extended the ban to almost all production, storage, and transport.

During the Sino-Japanese War of 1937–1945 and in World War II, Unit 731 of the Imperial Japanese Army conducted experiments on thousands of prisoners, mostly Chinese. In certain military campaigns, the Japanese used biological weapons on soldiers and civilians. The Japanese secretly fed their Chinese prisoners poisoned food. They also contaminated the water. Estimates suggest that more than five hundred thousand people died, due to the bad food and also to plague and cholera outbreaks.

Suspicious of reported biological weapons development in Germany and Japan, the United States, the United Kingdom, and Canada initiated a biological weapons development program in 1941 that resulted in the weaponization of anthrax, brucellosis, and botulinum toxin. The center for U.S. military biological weapons research was Fort Detrick, Maryland. Biological weapons research was also conducted at Dugway Proving Grounds in Utah. Research carried out in the United Kingdom during World War II left Gruinard Island in Scotland contaminated with anthrax for the next forty-eight years.

When biological and chemical weapons became too old, they sometimes needed to be disposed of. Many NATO nations used the U.S. chemical weapons disposal facility on Johnston Atoll, located in the middle of the Pacific.

Despite having signed the 1972 treaty, the Soviet Union continued research and production of offensive biological weapons in a program called Biopreparat. The United States was unaware of this program until Dr. Kanatjan Alibekov, the deputy director of Biopreparat, defected in 1992.

During the Cold War era, considerable research on biological warfare was performed by the United States, the Soviet Union, and other major countries. It is generally thought that such weapons were never used. Through 1998, more than 135 countries had signed the Biological Weapons Defense Bill of 1972. Still, not everyone felt that the world was any safer from a bioweapon attack.

In 1986, the U.S. government spent $42 million on developing defenses against infectious diseases and toxins, ten times more money than was spent in 1981. The money went to twenty-four U.S. universities, in hopes of developing strains of anthrax, Rift Valley fever, Japanese encephalitis, tularemia, shigella, botulin, and Q fever.

The United States maintained a stated national policy of never using biological weapons under any circumstances, as stated by President Nixon in November 1969.

At present, several countries are developing biological warfare programs. According to the U.S. Defense Department, these countries include Russia, Israel, China, Iran, Libya, Syria, and North Korea. The perfect characteristics of biological weapons is that they are highly infective, they have a high potency, they can be delivered as aerosols, and vaccines are unavailable for the victims.

The biggest problem with biological warfare is that while biological weapons can be made quickly and easily, the delivery of the weapons themselves is much more difficult. Take anthrax, which has been featured in the news a great deal since 9/11. Anthrax is considered an excellent biological weapon. Anthrax forms hardy spores, perfect for dispersal using aerosols. Plus, pneumonic (lung) infections of anthrax usually don't cause secondary infections in other people. Thus, the effect of the drug is usually confined to the target. A pneumonic anthrax infection starts with ordinary cold symptoms and quickly becomes lethal, with a fatality rate that is 80 percent or higher. One last fact to consider is that there is an anthrax vaccine, so soldiers using the drug as a weapon can be kept safe by taking antibiotics.

Conducting a major anthrax attack wouldn't be easy. A mass attack using anthrax would require the creation of aerosol particles of 1.5 to 5 micrometers. Too large and the aerosol germs would be filtered out by the respiratory system. Too small and the aerosol would be inhaled and exhaled. At this size, nonconductive powders tend to clump and cling because of electrostatic charges. This hinders dispersion. So the spores must be treated with silica to insulate and discharge the charges. Plus, the aerosol needs to be delivered so that rain and sun don't damage it and the human lung can still be infected. Still, despite these problems, lethal anthrax sprays have been produced and could be used in a deadly bioweapons attack on unsuspecting targets.

Other diseases considered for use as weapons or known to have already been used as weapons include anthrax, Ebola, bubonic plague, cholera, tularemia, brucellosis, Q fever, Machupo, coccidioidesmycosis, glanders, melioidosis, shigella, Rocky Mountain spotted fever, typhus, psitacosis, yellow fever, Japanese B encephalitis, Rift Valley fever, and smallpox. Naturally occurring poisons that can be used as weapons include ricin, SEB, botulism toxin, saxitoxin, and many mycotoxins.

Biological warfare can also specifically target plants to destroy crops or defoliate vegetation. The United States and Britain discovered plant growth regulators (herbicides) during World War II and initiated a herbicidal warfare program that was used in Malaya and Vietnam. Though herbicides are chemicals, they are often grouped with biological warfare as bioregulators in a similar manner as biotoxins.

During the Cold War, the United States developed an anticrop strategy that was aimed at using plant diseases (bioherbicides or mycoherbicides) to destroy enemy agriculture. The CIA believed that the destruction of the Russian wheat and grain fields on a huge scale would slow down or stop any Russian attack during a war. Diseases such as wheat blast and rice blast were turned into aerial spray tanks and cluster bombs for delivery to enemy watersheds in agricultural regions to initiate plant epidemics. When the United States renounced its offensive biological warfare program in 1969 and 1970, the vast majority of its biological arsenal was composed of these plant diseases.

Attacking animals was another area of biological warfare intended to eliminate animal resources for transportation and food. In World War I, German agents were arrested while attempting to inoculate draft animals with anthrax. In World War II, the British considered tainting small feed cakes as a way to attack German cattle. The plan was to thin the herds to cause food shortages, but the idea was never put into effect. In the 1950s, the United States government experimented with hog cholera but never actually used it as a biological weapon.

A plague that affects only animals might seem like an inconvenience to most people, but such a disease could wreak havoc on a nation's economy.

In 2001, an occurrence of foot-and-mouth disease in Great Britain resulted in the slaughtering and burning of 7 million sheep, pigs, and cows. Efforts to contain the disease, which could be transmitted on people's shoes or clothing, resulted in numerous sporting and social events being canceled. Tourism dropped sharply during the epidemic, and the government moved local and general elections to a date a month later. The cost of the disease to the British economy was estimated at more than $15 billion.

Of all the threats mentioned when discussing global terrorism, the use of biological weapons is without question the most horrifying. Cars, trains, and airlines have drawn the entire world closer together. A pandemic spreading from city to city, country to country, would be almost impossible to stop, especially if the disease had a three- or four-day incubation period. The first four hundred pages of *The Stand* makes it clear that no one is safe, no matter where he or she lives, when a major epidemic hits. The plague in *The Stand* could happen. It's not even a question of how, but mostly a question of when.

The Secret History of AIDS

One reason *The Stand* remains so popular is that many people don't believe that the book is entirely fiction. There's a dark undercurrent of belief that is commonplace among the poor and disenfranchised, particularly those who live outside the United States, that the U.S. government has already created and used a biological weapon in Africa. They feel certain that the "haves" of the world have decided to wipe out the "have-nots" using a man-made virus. The plague is called AIDS. It's an incredible scenario, right out of the pages of a thriller novel, but it's a story that many people believe is true. It's *The Stand* come to life, with us as the villains.

The theory was presented as fact by Dr. Robert Strecker in the 1991 "Strecker Memorandum," which now can be found all over the Internet and has even been made into a ninety-minute video. The plot goes like this: In a speech given at Notre Dame University in 1969, then secretary of defense Robert McNamara described the population explosion in Africa as one of the greatest dangers threatening the modern world. At the time, his thinking was in line with many of the top scientists of the day, who painted a bleak picture of millions starving to death in the 1980s and 1990s due to lack of food. There were only two ways to prevent this disaster, which would result in wars over food and grain throughout all of Africa and Asia. The two solutions were to drastically cut the birthrate of the poor countries or somehow drastically cut the population. The second method was decided upon by the politicians in Washington, and research was begun to develop a man-made plague that would primarily kill most of the people in Africa.

WHO, evidently feeling that the overpopulation problem had to be dealt with, issued a bulletin stating that "[a]n attempt should be made to see if viruses can in fact exert selective effects on immune function. The possibility should be looked into that the immune response to the virus itself may be impaired if the infecting virus damages, more or less selectively, the cell responding to the virus."[15] In other words, scientists should experiment with viruses that would destroy the immune systems of humans—which pretty much sums up what AIDS does.

Over the next few years, Strecker claimed that WHO scientists, along with experts from the National Cancer Institute, worked on the AIDS virus at Fort Detrick, Maryland, and finally achieved success in 1974. AIDS, Strecker claimed, was not related to a monkey virus but instead was a combination of two deadly retroviruses, bovine leukemia virus and sheep visna virus, injected into human tissue cultures.

With the virus completed, the next thing was to distribute it. This was done, according to the theory, by secretly contaminating the WHO smallpox vaccine used in Africa. Millions upon millions of smallpox vaccines were injected into unsuspecting African men, women, and children. The vaccine was also used in Haiti, Brazil, and Japan, thus resulting in huge outbreaks of AIDS in those countries as well.

The spread of AIDS to the United States was the result of more government biological warfare. In this case, the AIDS virus was put into the hepatitis-B vaccine developed by the government in 1982. Since hepatitis B was very common in homosexuals, mixing the AIDS virus with the vaccine was a sure way of infecting most of the homosexual community in the United States.

As mentioned, the "Strecker Memorandum" and videotape read and sound like the worst paranoid fiction imaginable. Yet many people, both in the United States and in the rest of the world, believe its every word. Wangari Maatha of Kenya, the first African woman ever to win the Nobel Peace Prize, in an interview with a Kenyan paper, said that she was certain AIDS was a man-made virus and was deliberately created as a weapon of biological warfare.[16] Like many stories on the Internet, the "Strecker Memorandum" shows no sign of disappearing anytime soon.

Is the document or the video true? Obviously, bits and pieces of legitimate documents are woven into the memorandum, giving it the look and feel of authenticity; however, all of its major points have been refuted not once but many times. HIV has nothing to do with sheep or bovine retroviruses. Instead, it closely resembles SIV, a disease infecting monkeys in Africa. The beginnings of AIDS occurred in Africa in the 1950s when poor hunters evidently killed and ate tainted monkeys. The virus was transmitted by the monkey flesh and blood into the men's intestines where it transformed into a human virus, thus resulting in the earliest cases of AIDS being reported back in the late 1950s in remote sections of Africa.

Numerous Web sites have analyzed the so-called undisputable facts of the "Strecker Memorandum," and in every case, the facts are more suppositions than basic truths. As pointed out by more than one government scientist, the skill necessary for creating the AIDS virus was just not available to researchers in 1974. The "Strecker Memorandum" tells a fascinating story of governments conducting biological warfare on their own citizens, but when it comes to the so-called secret truth behind the memorandum, the "Strecker Memorandum" is no more real than *The Stand* is.

5

UP THE DIMENSIONS WITH STEPHEN KING

The Dark Tower I: The Gunslinger • Insomnia

The man in black fled across the desert,
and the gunslinger followed.

—*The Dark Tower I: The Gunslinger*

The Dark Tower books began when college student Stephen King encountered Robert Browning's poem "Childe Roland" and soon afterward watched the spaghetti Western *The Good, the Bad, and the Ugly*, the third chapter in the Man with No Name adventures starring Clint Eastwood. Those two inspirations and a big stack of unused green paper started King thinking, and from his mind emerged the first line that began an epic: "The man in black fled across the desert, and the gunslinger followed."

It took King twelve years to finish the series of interconnected novelettes that formed *The Gunslinger*, the first book of his epic seven-volume series The Dark Tower. During that time, he went from being a college student to one of the most popular authors of the late twentieth century.

The World of the Dark Tower

The Gunslinger stories tell of an alternate universe where the last gunfighter, Roland Deschain, pursues the mysterious Walter O'Dim, the Man in Black, across an eerie landscape similar to and yet different from Earth, called the All-World. It is Walter who can guide Roland to his final destination, the Dark Tower, the place that defines all reality. Accompanying Roland on the first part of his odyssey is a young boy named Jake, who comes from our world in the 1970s.

In the series, an ancient race called the Old Ones once lived on All-World, but its members have since disappeared. No one is sure what happened to the Old Ones, but a combination of germ warfare and nuclear war offers a possible explanation. Many of the inhabitants of All-World believe that the being known as the Crimson King was behind the Old Ones' disappearance. The main consequence of the end of the Old Ones leaving is that the world is slowly coming apart, falling into ruins. Entire cities vanish, great wars are fought for no reason, and sometimes the sun rises from the north and sets in the east. Sometimes the sun seems not to set at all but instead just disappears into haze. Time runs at different speeds in All-World, and nothing that happens is permanent. This world is dying. In King's words, it has "moved on."

Roland's mission is to find the mythical and mystical structure known as the Dark Tower. This place is said to be the nexus of the universe, the place where all realities from all worlds and dimensions intersect. When we first meet Roland, his reasons for searching for the Dark Tower aren't very clear, although he is obsessed with his mission. As the epic adventure continues, Roland's main goal, to save the universe from destruction by the Crimson King, becomes apparent.

All-World is also called the Keystone Tower, as it is the only world in the multiverse (the universe of all universes) that contains the Dark Tower in its actual physical state. In other worlds in other universes, the Tower is present but inaccessible. Only by going through All-World can someone actually enter the Dark Tower.

The Crimson King wants to destroy the Dark Tower, which would result in the destruction of the universe, leaving him alone to rule the primal darkness from which all things arose. The Crimson King is trapped on a balcony of one of the floors of the Tower.

It's worth mentioning that Stephen King wasn't the only writer to use this concept of a multiverse as a foundation for his fictional universe. In the 1960s, the British author Michael Moorcock began a series of fantasy adventures about an albino antihero named Elric. The Elric series were also published first as a series of novelettes, which were later published in book form as novels. In Moorcock's series, Elric is portrayed as one incarnation of a being known as the Eternal Champion, a heroic figure who fights evil in numerous parallel worlds, including modern-day Earth. Moorcock wrote dozens of novels describing the adventures of the Eternal Champion in the multiverse, including several involving a mysterious Vanishing Tower.

King, in response to an e-mail sent February 10, 2006, from Dark Tower scholar Bev Vincent, stated that he never read any of Moorcock's work. In an e-mail dated November 21, 2001, published on the Web site Moorcock's Miscellany, Michael Moorcock mentioned that he never read any of King's work. Thus, two giants of the fantasy and horror field both wrote multibook sagas about iconic, lone heroes wandering through a multiverse without once realizing the kinship between their fiction.

The Dark Tower series touches upon actions and events important to many of Stephen King's novels. He describes the universe of the Dark Tower as existing inside a single molecule that is part of a stem of grass that lies in a field of grass that has just been cut by a lawn mower. The grass is lying on the field, dying, and all of the life inside the one particular blade of grass, including the world of the Dark Tower, is dying as well.

Oddly enough, despite the huge size of the Dark Tower series, the nature of reality as espoused in the epic and that underlies many of King's other novels is never fully explained in the seven novels that make up the story. That's left to *Insomnia* (1994). In that novel, Stephen King first explains the true nature of the Dark Tower, introduces the Crimson King, and discusses in depth the meaning of life.

Sleepless in Maine

Insomnia tells the story of Ralph Roberts, a retired widower who, after the death of his wife, begins to suffer from lack of sleep. Ralph has no problem falling asleep; he just wakes up earlier and earlier each night until he finds himself napping for only two or three hours each evening. A stubborn man, Ralph tries every possible remedy but refuses to visit his family doctor since the physician misdiagnosed Ralph's wife's cancer. As his insomnia grows worse and worse, Ralph starts seeing things that are invisible to his friends and neighbors in the mystical town of Derry, Maine.

Mostly, Ralph sees different-colored balloons that seem to sprout from the heads of the town's inhabitants. It doesn't take him long before he figures out that these balloons, which end in lines of light like strings, are the life-force auras of people (as well as of animals). After observing the auras carefully, Ralph comes to believe that they reflect the health and well-being of the people and the animals that emit them. Still, he's not too concerned about his visions until he notices three short figures in white coats roaming around Derry, unseen by everyone but him. These "little bald doctors," as he calls them, are very aware of the life-force auras. Two of the little beings seem benevolent, but the third doctor, armed with a pair of rusted scissors, strikes Ralph as being evil.

When Ralph discovers that his long-time friend and fellow senior citizen Lois Chasse also can see auras, he teams up with her to try to find out the truth about the three little bald doctors, whom he dubs Clotho, Lachesis, and Atropos, after the three Fates. When confronted by Ralph and Lois, Clotho and Lachesis prove to be more than willing to talk about their purpose.

According to the two friendly Fates, four constants exist in the universe: Life, Death, the Purpose, and the Random. The three Fates are all agents of Death. The two that Ralph first meets, however, are also agents of the Purpose, while the third, Atropos, the one with the rusted scissors, is an agent of the Random.

Humans (called "short-timers" by the Fates) live on the bottom level of the Tower and normally can perceive only the first two levels. The Fates (calling themselves "long-timers") live on a higher level. The artificially induced insomnia that plagues both senior citizens is what enables them not only to see higher levels of reality but also to travel in them. The Tower has many levels, much higher than even the Fates know. Nor do they know what is at the top. They do mention that on the upper levels live beings of incredible power that they call the Higher Purpose and the Higher Random.

Needless to say, Clotho and Lachesis aren't telling Ralph and Lois the secrets of the universe just because they like the two senior citizens. The two Fates want the old man and his friend to thwart a plot by the Higher Random to use Atropos to disrupt the balance between the Purpose and the Random. How Ralph and Lois accomplish this, traveling up and down through various stages of reality, and their rewards and penalties for helping two of the Fates and defying the third make up the rest of the book. In the end, we are told: "Upon all the levels of the universe, matters both Random and Purposeful resumed their ordained courses. Worlds which had trembled for a moment in their orbits now steadied, and in one of those worlds, in a desert that was the apotheosis of all deserts, a man named Roland turned over in his bedroll and slept easily once again beneath the alien constellations."[1]

What's made clear by *Insomnia* is that the overlapping realities of the Dark Tower series are not alternate worlds where a historical event took a wrong turn. The layered realities of the Dark Tower, one rising above another in *Insomnia* like a skyscraper into the darkness, are in no way the same as the alien landscape of *From a Buick 8*. Nor do the horrid monsters from "The Mist" come from a floor in the Tower. The many-layered realities of Stephen King's Dark Tower are completely different from the alien Earths of Hugh Everett III's many worlds interpretation of quantum mechanics (for a thorough discussion of parallel worlds, be sure to check out chapter 7 of this book).

That leads us to ask the question whether there's any scientific theory that somehow explains the levels of reality stacked one upon another in the Dark Tower.

We think so, although, because of technical difficulties, proving anything about a twenty-first-century theory with twentieth-century tools, we can't say for sure that it's true: at least, not yet. The hypothesis that we believe best applies to the universe of the Dark Tower is known to modern scientists as M-theory. The name of the theory comes from its discoverer, Ed Witten, who is first to admit that the M stands for "magic," "mystery," or "matrix," depending on one's taste. We like to think the M stands for "maybe."

A Basic Course in Four-Dimensional Geometry

To understand M-theory, we first need to examine dimensions one through four and see what we can deduce from our knowledge of them. To understand eleven dimensions (discussed later in this chapter), we must make sure we fully understand the first four. Then, hopefully, we can observe how that information applies to the Gunslinger saga.

A point is an object with zero dimensions. It does not extend into space and it has no properties. It has no length, no width, and no height.

An object with one dimension is a line. A line has a beginning and an end if it is finite, or it can stretch out to infinity if it has no end. A line has no thickness. Using the various systems of measurement, a finite line can be described as being a certain length.

If we take one line and place it at a right angle to another line, we end up with a two-dimensional object, a plane: L. Two lines at right angles form a plane that has length and breadth but no thickness.

Imagine a plane as a piece of paper but with absolutely no depth.

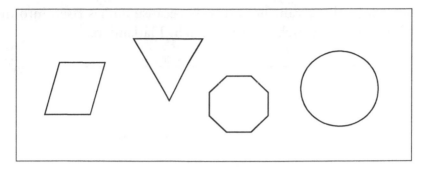

Figure 1. Objects on a plane.

Objects on a plane can have a number of different forms. They can be circles, triangles, squares, rectangles, or any combination of such objects. The one property that they all have in common is that they have only two dimensions, length and breadth (see figure 1).

Now, imagine an intelligent being living on this sheet of paper of no thickness. Because the world is absolutely flat, we might think of it as a flat-land, which just happens to be the title of a famous book dealing with dimensions, *Flatland*, by Edward Abbott (1884). In Flatland, people are circles, squares, rectangles, and triangles. Social status is determined by a person's shape. One extremely important fact is that because these Flatlanders are two dimensional, they cannot see anything that is above or below them. They can perceive things only on the flat plane of their existence. In fact, it is impossible for them to even imagine what anything above them would mean because they would have to visualize something that they could not see—a higher dimension.

In *Flatland*, the narrator is visited by a being from another dimension, a sphere (from Sphereland). The sphere has three dimensions, length, width, and height; however, our Flatlander hero can see only the part of the sphere that intersects Flatland. Thus, the sphere appears as a circle in Flatland (see figure 2). Moreover, as the sphere sinks into the two-dimensional space of Flatland, he appears to grow into a larger and larger circle. After he passes more than halfway through Flatland, he then appears to grow smaller and smaller until he turns into a point instead of a circle. When he leaves Flatland, he vanishes entirely because he has risen into the third dimension, which can't be seen by Flatlanders.

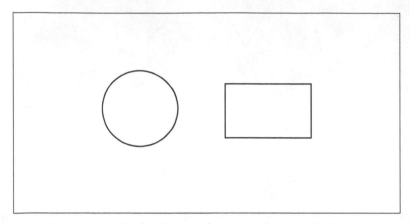

Figure 2. A Flatlander confronts a sphere in Flatland.

Needless to say, the Flatlander has a difficult time convincing any of the other inhabitants of the existence of a third dimension, much less beings that come from that higher dimension. The Flatlander is judged crazy and locked up.

Several points not discussed in *Flatland* but touched upon in a sequel written many years later (*Sphereland*, by Dionys Burger, 1965) are worth noting. The sphere (or any other inhabitant of the three-dimensional world) can move entirely in the third dimension. Being above Flatland, he can therefore see the inside of any of the two-dimensional inhabitants of that world. More frightening (in a horror-story manner), he can actually materialize inside any of the Flatlanders, with his body appearing inside theirs by traveling through the third dimension (see figure 3).

Equally important, while the Flatlander can perceive only two dimensions, his flat world could quite easily be curved in three dimensions. Imagine that the piece of flat paper we call Flatland is made out of rubber; then picture it wrapped around a globe. The inhabitants of Flatland, who can perceive only two dimensions, would not know that their world is curved in three dimensions, but looking into the distance, they would be able to see only so far.

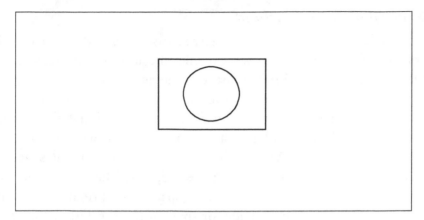

Figure 3. A sphere materializes inside a Flatlander.

When traveling in a straight line on their world, they would continue forward and yet end up back where they started. Plus, if they somehow were able to travel in the third dimension (of depth), they could go from one place in Flatland to another without traveling across the world. Yet they would arrive where they were going by traveling a shorter distance through the higher dimension.

Leaving Flatland, we move from two dimensions to three. Three dimensions are defined as three straight lines all at right angles to one another (figure 4). We live in a three-dimensional world of length, width, and height.

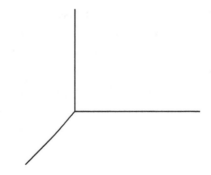

Figure 4. Three dimensions.

Extrapolating on Flatland

Let's apply some of the lessons learned from *Flatland* to our world. First and foremost, while we can imagine a fourth dimension of measurement (if we don't consider time as part of our scenario), since our minds are limited by the three-dimensional boundaries of our existence, we cannot physically comprehend what the fourth dimension looks like. The best we can do is use what is called a dimensional analogy. We study how a two-dimensional space is related to a three-dimensional space and, from that, draw conclusions about how a three-dimensional space would relate to a four-dimensional space. (See earlier for some of those relationships.)

Using this analogy, we can conclude that a being in four dimensions would be able to enter a locked room without opening or closing a door. Or he would be able to remove the motor from a car without opening the hood. Or (in regard to a horror novel) remove a person's heart without making one cut. H. P. Lovecraft used hyperspace travel extremely effectively in his classic horror story "The Dreams in the Witch House" (*Weird Tales*, February 1933).

Four-dimensional beings would be visible to us only when they intersect our three-dimensional world. Thus, they might appear to be horrifying monstrosities because we would be seeing only part of their entire bodies. The science-fiction writer Donald Wandrei wrote an extremely effective story about such an occurrence titled "The Blinding Shadows" (*Astounding Science Fiction*, May 1934). Four-dimensional beings would change shape from instant to instant in our three-dimensional world, much as the sphere did in *Flatland*. Plus, they could go from place to place without being seen if they traveled entirely in the fourth dimension. As has been suggested in more than one science-fiction story, four-dimensional beings would seem like gods, or demons, to inhabitants of our three-dimensional world. They would have powers that would seem supernatural to us but would be perfectly normal to them—which is very much how the three beings called the Fates are described in *Insomnia*.

The word *hyper*, from the Greek meaning "above" or "more than," is often used in relation to the fourth dimension. A square in three dimensions is a cube. A cube in four dimensions is called a hypercube, or a tesseract. In a famous science-fiction story, "And He Built a Crooked House" (*Astounding SF*, February 1941), Robert A. Heinlein describes a group of people caught in a house built in the form of a tesseract.

Hyperspace usually refers to spaceships entering the fourth dimension. Using our dimensional analogy helps to make the concept of hyperspace, as used in science-fiction stories and television, comprehensible. Just as we stretched a two-dimensional world (Flatland) onto a globe, we can suppose that our three-dimensional universe is stretched across a four-dimensional sphere. In three dimensions, the fastest method of traveling between point A and point B is by taking the straight line connecting the points; however, since the universe is curved in four dimensions, the actual shortest distance between the two locations is through hyperspace.

Of course, since space is curved through the fourth dimension and we can't picture what that curve might look like, it's not 100 percent guaranteed that traveling through hyperspace will lead from point A to point B. Nor do we have any idea what hyperspace might look like and whether we would be able to find our way around it. Just as the dweller on Flatland could not describe or visualize Sphereland, we cannot visualize or describe the world of the fourth dimension. It's likely that the stars and the planets might not intrude into hyperspace, and it's also equally likely that some sort of fourth-dimensional objects might be there in their place.

It's quite possible that if we ever enter the fourth dimension, we won't be able to find our way out. Plus, we have no clear idea how to enter hyperspace from our three-dimensional world, nor do we have any idea, once we are in that region, how to navigate there or exit, which is why it remains science fiction.

Insomnia: *Visiting Other Dimensions*

In *Insomnia*, Ralph Roberts travels from a portable Port-O-San unit in the park in Derry through hyperspace to the cockpit of the plane Ed Deepneau is flying ten thousand feet over the city. That's where he engages Ed in a life-or-death battle to save thousands of Derry's citizens from being killed. Or, in terms that Ralph uses earlier in his adventures with the Fates, he travels up and down "narrow staircases festooned with cobwebs and doorways leading to God knows what" in the building known as the Dark Tower.

It's quite clear from reading *Insomnia* that the different levels of the Dark Tower are the different levels of reality. The first two floors are the levels of three-dimensional space, while the third and perhaps fourth floors are the levels of four-dimensional space. Humans live on the first two levels, while the Fates and similar creatures as those who appear in the Dark Tower novels live on the third and fourth floors. Humans, as described in *Insomnia*, are four-dimensional beings, but since people have only three-dimensional senses, they cannot see the fourth-dimensional auras that flare out from their bodies. Nor do these colored auras affect anyone else in the three-dimensional world as they are made of a substance that can be sensed only in the fourth dimension. But the Dark Tower, we are told by the Fates, stretches upward many floors—forcing us to ask several questions. Since humans can neither sense nor see the fourth dimension, does it really exist? And do any dimensions exist that are higher up than the fourth dimension? If so, can we prove they exist?

All of which brings us to Albert Einstein and the problem that he couldn't solve. The problem, as you might have guessed by now, involves M-theory.

Einstein and Heisenberg and King, Oh My!

Stephen King's Dark Tower series represents the author's effort to tie all (or as much) of his work as possible into one consistent universe, where the various individual stories are related by characters, scenery, and sometimes plot. Thus, a reader of *The Waste Lands* might come across a character from *The Stand* or *Salem's Lot* who suddenly appears halfway through the novel. Or he or she might discover that a song sung by a minor character in a King novel later turns out to have a much more important role in another part of the epic. In that manner, the Dark Tower series is central to all of King's

The notion of treating time and space as two parts of a unified continuum was first devised by Hermann Minkowski soon after Einstein announced his theory of special relativity. The concept of space-time was vital to Einstein's later work, the theory of general relativity, which explained gravity.

The theory of special relativity made clear one of the major problems with Newton's law of universal gravitation. In Newton's equation, one body exerted a gravitational pull on another. If the distance between the two objects changed, according to Newton's law, the gravitation felt by one of those objects would happen immediately. This implied that if the sun suddenly went out on a Newtonian Earth, people would know it instantly. Whereas if the sun went out on a Einsteinian Earth, people wouldn't find out until eight minutes later when the light and force waves hit the planet. Obviously, both facts couldn't be true.

Gravitation and Curved Space-Time

Using the theory of special relativity as his starting point, in 1915 Einstein developed what he called the theory of general relativity (also known as the general theory of relativity). Einstein's theory stated that gravitation was not a force but was actually a manifestation of curved space-time. When Einstein first presented his theory, it was regarded by most scientists as pure speculation. That was the case until 1919. That's when predictions that were made using Einstein's theory on how much light issuing from a star was deflected by the gravity of the sun were confirmed during a solar eclipse by Arthur Eddington. These observations were made on May 29, 1919, in two different locations, one in Sobral, Ceará, Brazil, and another on the island of Principe, off the west coast of Africa. This principle became known as gravitational lensing. When the observations were reported in the *Times of London*, Einstein became world famous and his theory was accepted as the cutting edge of physics. Since 1919, thousands more observations have proved that Einstein's theory of general relativity describing the motion of interstellar objects is incredibly accurate.

Einstein's theory is best understood using a dimensional analogy.

A scientist takes a large piece of rubber (or a trampoline) and stretches it over a four-cornered frame, thus forming a rectangle. According to Einstein, this rectangle represents three-dimensional space. The scientist then takes a bowling ball and drops it into the center of the stretched rubber rectangle. The heavy weight of the bowling ball makes a deep indentation in the rubber sheet, as well as curving the surrounding area of the sheet. The bowling ball represents the sun. When a marble is sent spinning around the bowling ball, it curves around the ball in a steady orbit due to the warped nature of the rubber sheet. The marble, of course, represents Earth or any other planet. The final step in the analogy is to imagine that the two-dimensional sheet is three-dimensional space, which implies that the sun warps space-time in four dimensions. Einstein's theory showed that gravity was related to mass and that the warping of the fabric of the universe is gravity.

Einstein's theory of general relativity states, in its most reduced terms, that mass and energy curve space and time in the fourth dimension. Einstein also provides the equation that actually determined the relationship, which he calls the field equation. The field equation is a tensor equation relating a set of symmetrical 4×4 tensors.[3] Thus, we have confirmed beyond the shadow of a doubt that the fourth dimension exists and most likely intersects the fourth floor of the Dark Tower.

The greatest challenge to the theory of general relativity came from Einstein himself, because he was interested not only in incredibly large objects but also in incredibly tiny ones. Like Isaac Newton before him, Einstein changed our views of the very large as well as of the very small.

Quantum Mechanics and General Relativity

In 1905, Einstein published a paper titled "On a Heuristic Viewpoint Concerning the Production and Transformation of Light." He argued that "energy quanta," which later became known as photons, were real and that they could be used to explain the photoelectric effect. Up until that time, scientist believed that light was a wave. In 1921, Einstein was awarded the Nobel Prize for this paper, although most people at the time felt that he was given the prize for all of his incredible discoveries.

In 1905, Einstein published a paper called "On the Motion—Required by the Molecular Kinetic Theory of Heat—of Small Particles Suspended in a Stationary Liquid." This study discussed Brownian motion and provided actual visible evidence of the existence of atoms. Before Einstein, not all scientists believed that atoms actually existed. This paper proved that they did and offered an actual method of counting them, using a microscope.

Einstein's research paper on the photoelectric effect showed that light possesses particlelike properties; however, it had been demonstrated in the nineteenth century that light possesses wavelike properties. Scientists were thus forced to conclude that light possesses the properties of both particles and waves. Seeking to understand this strange behavior, scientists, including Max Planck, Niels Bohr, and Werner Heisenberg (along with numerous others), developed a new branch of physics that they named quantum mechanics. Quantum mechanics dealt with atomic and subatomic systems and provided accurate and very precise descriptions for what occurred on these levels, something that could not be done using classical physics. Newton's laws and classical electromagnetism, for example, when applied to atoms, had electrons crashing inward and colliding with the nucleus.

Quantum mechanics was originally developed to explain the atom and especially the orbits and the positions of the electrons in atoms. Thus, quantum theory provided an explanation for the electron's staying in its orbit, an answer that couldn't be provided by Newton's laws of motion and by classical electromagnetism. Quantum mechanics uses complex number wave functions to explain these effects. The solutions were related to normal physics mostly due to the use of probability in solving the equations. Probability in quantum mechanics involved how likely it was to find a particle—say, an electron—in a certain region around the nucleus at a certain time.

The problem of applying general relativity to such small objects is that it doesn't work. Instead, the Heisenberg uncertainty principle governs the location of the particle. The principle says that it's impossible to know the exact position of a subatomic particle, so instead of saying where it is, an entire range of possible places is used, described by a probability distribution. Therefore, quantum mechanics, translated to Newton's equally deterministic description, leads to a probabilistic description of nature.

In the mid-1920s, as the new theory of quantum mechanics grew in popularity, Einstein found himself at odds with what became known as the Copenhagen interpretation of the new equations of quantum mechanics, because the solutions were a probabilistic, nonvisualizable account of physical behavior. Einstein wanted a more complete answer, a solution that was more deterministic, more exact. He didn't like a range of possible answers.

In a 1926 letter to Max Born, Einstein made a remark that is now famous: "Quantum mechanics is certainly imposing. But an inner voice tells me it is not yet the real thing. The theory says a lot, but does not really bring us any closer to the secret of the Old One. I, at any rate, am convinced that He does not throw dice."

The Old One whom Einstein refers to, the one who does not throw dice, is, of course, God. Einstein found himself confronted by a major conflict in physics. With the theory of general relativity, he had discovered the laws of physics that described the way the universe worked. With the push given by his 1905 paper on light, he had helped to create quantum mechanics, the physics that described the way atomic and subatomic particles worked. According to his personal beliefs in the order of the universe, the two theories should have been in harmony, with one reducing to an equivalent of the other. Instead, the two great theories contradicted each other. The theory of general relativity was a deterministic theory, and quantum mechanics was essentially an indeterministic theory. Albert Einstein, Boris Podolsky, and Nathan Rosen (collectively known as EPR) wanted to bring "elements of reality" to quantum mechanics to make it deterministic. They postulated the existence of hidden variables that, if only they could be measured, would make quantum mechanics deterministic. Professor John Howell of the University of Rochester, who has been doing experiments in quantum mechanics and the EPR paradox, stated in an interview with the authors, "Experiments to date have pointed to the fact that quantum mechanics seems to be indeterministic."

General relativity relied primarily on the force of gravity, while quantum mechanics relied on the other three fundamental forces in the universe: the weak nuclear force, the strong nuclear force, and the electromagnetic force. Try to solve a problem concerning the speed of galaxies by using quantum mechanics, and the answer is incomprehensible. Try to solve a problem regarding the motion of an electron using general relativity, and the answer most likely is infinity. Neither theory worked in the other's domain. It was a problem that drove the greatest minds of the twentieth century bonkers. Plus, it wasted a lot of their valuable time, which might have been better used working on other projects.

Needless to say, we can't ignore or abandon either theory. The theory of general relativity is an extremely useful model of gravitation and cosmology that has passed every unambiguous test it has been subjected to thus far, both observationally and experimentally. Since its inception, it has predicted numerous events in space, including the discovery of black holes. It is therefore almost universally accepted by the scientific community as truth.

On the other hand, quantum mechanics helps us to understand how individual atoms combine to form chemicals. The application of quantum mechanics to chemistry is known as quantum chemistry. Quantum mechanics provides quantitative insight into chemical bonding processes by showing which molecules are energetically favorable to which others and by approximately how much. Most of the calculations performed in computational chemistry rely on quantum mechanics.

Also, a great deal of modern technology operates on a scale where quantum effects are significant. Some examples include the laser, the transistor, the electron microscope, and magnetic resonance imaging. The study of semiconductors led to the invention of the diode and the transistor, both of which are indispensable for modern electronics.

Both theories have been demonstrated to work perfectly again and again over the last seventy-five years. Yet, like Albert Einstein, our mathematical intuition tells us that there cannot be two entirely different theories that explain how the universe, from atoms to galaxies, works. Like the Dark Tower, there can be only one.

The greatest problem in modern physics is to somehow find a theory that unifies everything we know about quantum mechanics and general relativity. That's the question that has been driving physicists to drink for the last seventy-five years. Only recently do they think they may have a glimpse of the answer, but proving that the theory is true requires mathematics and physics we don't yet possess. The solution involves the shortest possible objects in the universe and, as we mentioned before, is called M-theory. For anyone who has been keeping track, it's also sometimes referred to as E-theory, where the E stands for "everything." It's the one ring that binds them all, the top room of the Dark Tower. If only we could fully understand what it's all about.

The Theory of Everything

As stated earlier, the central conflict of modern physics was that general relativity conflicted with quantum mechanics. Attempts to merge the two theories resulted in nonsensical answers, which made it clear to scientists that they needed a new theory that somehow combined the principles of both. Some physicists were willing to accept that there were two theories: general relativity for dealing with outer space and quantum mechanics for dealing with particle physics—the study of atoms and atomic particles. Unfortunately, there were certain problems that involved both subjects at the same time. One such problem was the big bang that began our universe. Not being able to understand how the universe was created was a constant irritation to physicists, so the search for Einstein's unified field theory continued.

Meanwhile, confronted by a problem they couldn't solve, physicists spent much of their time trying to comprehend the two theories they already possessed. Investigations of general relativity resulted in a better understanding of the cosmos and demonstrated that Einstein's equations were incredibly accurate. In quantum mechanics, however, the results were more astonishing.

Over the course of decades, physicists discovered that the structure of the atom was much more complex than they had realized—and that the forces acting between atomic particles were three of the four fundamental forces of the universe. Putting all these facts together in the 1970s, physicists came up with what became known as the standard model of quantum mechanics.

Quirky Quarks and Other Fundamental Particles

The standard model stated that there were two types of fundamental particles: fermions and bosons. They were called fundamental because they couldn't be broken into smaller components. These incredibly small particles were treated by scientists as zero-dimensional points in their calculations.

There were two types of fermions: quarks and leptons. The concept of quarks was first postulated in 1961, independently, by Murray Gell-Mann and Kazuhiko Nishijima. It was Gell-Mann who named these particles *quarks*, taking the name from a line in James Joyce's book *Finnegan's Wake*, "Three quarks for Muster Mark."[4]

Quarks exist only in groups of two or three. They combine to make bigger and more stable particles. The force holding them together is known as the *color force*, which used to be called the strong nuclear force. Particles that are a combination of a quark and an anti-quark are called *mesons* and are usually the result of high-energy collisions between particles. Three quark particles are called *baryons*, the most common of which are protons and neutrons. Protons and neutrons combine to form the nucleus of an atom. The strong force holding protons and neutrons together is a by-product of the color force holding the quarks together. As Professor Arie Bodek, the chair of the Department of Physics and Astronomy at the University of Rochester, stated in an interview with the authors, "The relationship between the force binding the proton and neutron together to the color force between the quarks is analogous to the force binding two atoms together to form a molecule, which is a by-product of the electric force." Bodek, whose doctoral thesis provided the evidence for the existence of the quark, uses electrons and neutrino beams to probe the distribution of quarks inside the proton and the neutron.

There are six types of quarks: up, down, strange, charm, bottom, and top. Fortunately, for our discussion, the type of quarks involved doesn't matter.

Leptons are indestructible. Unlike quarks, they don't combine to make bigger particles. Leptons include electrons, positrons, neutrinos, and muons.

Electrons are attracted to protons because they have opposite electrical charges. This attraction is known as electromagnetic force. The position of the electron inside an atom is determined by its energy. Electromagnetic force is infinite in range, but it works only between charged particles.

Gravity affects everything in the universe, from atoms to stars. Gravity works by pulling everything together. Gravity is infinite but is the weakest force. Gravity is the one force not directly involved with the standard model.

Thus, the four fundamental forces of nature are gravity, electromagnetism, and the weak and the strong nuclear forces. The strongest of these forces is the strong nuclear force; then comes the weak nuclear force, followed by electromagnetic force, and finally the weakest force, the gravitational force.

Gauge bosons are defined as boson particles that act as carriers of the fundamental forces from one particle to another. There are three known gauge bosons: photons, vector bosons (W and Z bosons), and gluons. When a gauge boson collides with a particle, its energy is absorbed and it ceases to exist. Photons transfer energy for the electromagnetic force. W and Z bosons transfer energy for the weak force, and gluons transfer energy for the strong force. A fourth type of gauge boson, the graviton, which transfers energy for the gravitational force, has been postulated but never observed (or confirmed to exist).

Veneziano and String Theory

Nearly all experiments conducted in particle physics of the three forces discussed in the standard model agreed with the information predicted using the model. The problem with the standard model, however, was that it couldn't lead to a unified field theory because it didn't deal with gravity. So, while scientists had a better understanding of subatomic particles, they still couldn't connect quantum mechanics with general relativity—until Gabriele Veneziano, a physicist in Europe studying the strong nuclear force in 1968, noticed a strange coincidence.

Veneziano discovered that a fairly obscure mathematical formula by the Swiss mathematician Leonhard Euler, called the Euler-beta function, described many of the properties of strongly interacting particles. Physicists from around the world started to study Veneziano's discovery. It soon became clear that the Euler-beta function described the strong nuclear force between particles, but no one understood why. That changed in 1970 when three physicists realized that if, instead of describing particles as points, they were treated as tiny, vibrating one-dimensional strings, then the strong nuclear force acting on them was perfectly detailed by Euler's formula. The size of the strings was estimated to be somewhere around the planck length (about 1.6×10^{-35} meters). This new field, which the scientists dubbed *string theory*, explored the implications of strings that were open or closed. The scientists further reasoned that since the strings were incredibly small, they would still resemble points and would thus be consistent with standard experiments. Plus, the idea would explain why there were no single quarks, since it's impossible to have a string with only one end.

This first breakthrough concerning strings attracted some notice, but before long, experiments in the early 1970s produced results that seemed to contradict many of the theory's predictions. The problem was that the vibrations of the strings produced not only properties that corresponded with gluons but with other patterns as well. These other patterns seemed to have no meaning until two scientists, John Schwarz of Caltech and Joel Scherk of Ecole Normale Superieure, demonstrated that the other patterns corresponded with the theorized properties of gravitons. Though gravitons had never been found, scientists had predicted their existence and properties. It seemed that a theory had finally been formulated that tied quantum mechanics with general relativity.

"I was immediately convinced this was worth devoting my life to," Dr. Schwarz said in a 2004 interview. "It's been my life work ever since."[5]

Still, most scientists felt there were too many problems between string theory and quantum mechanics. According to Schwarz, "Our work was universally ignored."[6] The theory was disregarded by nearly everyone—except Schwarz.

In 1984, Schwarz, working with Michael Green of Queen Mary College (Scherk died in 1980), published a groundbreaking paper on string theory. The two physicists showed that the conflicts between string theory and quantum mechanics could be resolved. More important, they demonstrated that string theory didn't merely account for gravitons and gluons but offered explanations for all of the four fundamental forces of nature and all the matter in existence. According to string theory, the basic components of matter and energy, fermions and bosons, were both strings, the only difference being in the way they vibrated. Suddenly, string theory was the hottest theory in physics—thanks in part to some help from Theodor Kaluza and Oskar Klein, as well as Eugenio Calabi and Shing-Tung Yau.

Back in 1919, the Polish mathematician Theodor Kaluza proposed that the existence of a fourth spatial dimension might allow the linking of general relativity and electromagnetic theory. Kaluza, working with the Swedish mathematician Oskar Klein, suggested that space consisted of both extended and curled-up dimensions. The extended dimensions were the three spatial dimensions of length, width, and height. The curled-up dimension was found deep within the extended dimensions and was thought of as a circle. Experiments later showed that Kaluza and Klein's curled-up dimension wasn't the solution to the unified field theory. Still, their work did wonders for string theory.

The mathematics used in superstring theory required at least ten dimensions. In other words, to solve the equations that described string theory and for those equations to somehow connect general relativity to quantum mechanics, they needed to use more than four dimensions of time and space. The only way to explain the nature of particles and to unify the four fundamental forces of nature required equations set in ten distinct dimensions. Those extra dimensions, string physicists decided, were wrapped up in the curled-up space first described by the team of Kaluza and Klein.

The way that string scientists extended the curled-up space to include more dimensions was by replacing the circles described by Kaluza and Klein with spheres. Thus, instead of adding only one dimension, a sphere added two more, the inside and the outside of the sphere, and a third, the space within the sphere. Added to the original three dimensions of measurement, scientists had six dimensions.

Years before string theory, a pair of mathematicians, Eugenio Calabi of the University of Pennsylvania and Shing-Tung Yau of Harvard University, wrote a paper describing six-dimensional geometrical shapes. String theory scientists familiar with these Calabi-Yau shapes merely substituted the six-dimensional shapes in the curled-up dimension of Kaluza and Klein. By doing this, physicists ended up with a space having ten dimensions—the three dimensions of normal space, the six of the Calabi-Yau shapes, and one dimension of time. String theory mathematics was possible, and the field expanded like a star going nova.

During the period 1984 through 1986, more than a thousand papers were published on string theory. These studies showed that the numerous features of the standard model could be derived from the new string theory. Not only that, but the explanation for many of those facts was much fuller and more convincing. The codiscoverer Michael Green put it best: "The moment you encounter string theory and realize that almost all of the major developments in physics over the last hundred years emerge—and emerge with such elegance—from such a simple starting point, you realize that this incredibly compelling theory is in a class by its own."[7]

Still, despite Green's enthusiasm, string theory did not overwhelm physics—mostly because the equations governing string theory proved to be so difficult, not only to solve, but just to determine, that scientists working in the field were forced to use approximations to replace them. After years and years of working with approximate equations of incredible complexity, many scientists lost patience with string theory and returned to their previous studies.

Another, entirely different but equally serious, problem plagued string theory. In the mid-1970s, physicists working entirely independently of string theory developed the concept of supersymmetry, the notion that symmetry applied to elementary particles. According to supersymmetry theory, every boson had a corresponding fermion partner, and every fermion had a corresponding boson partner. Then, in the 1980s, physicists realized that supersymmetry had to be a part of string theory if the theory was to cover both matter and energy. This modification of string theory led to it becoming known as superstring theory. Yet when physicists combined supersymmetry with string theory, they discovered that the merging of the two theories could be done in five different ways, with each new theory markedly different from the others. Thus, from 1985 on, scientists found that they had five superstring theories to investigate, not one.

For a time, the hope was that these five theories would prove to be the same principle, merely stated in a different manner; however, research on each theory convinced the scientists that all five theories were different. The one thing they did share was the fact that they required space to have ten dimensions for the theory to work.

It wasn't until a 1995 conference at the University of Southern California called "Strings 1995" that a solution to the five string theories was proposed. Dr. Edward Witten, of the Institute for Advanced Study in Princeton, New Jersey, declared that the five different string theories, each of which required ten dimensions to function properly, were all different manifestations of one unifying theory in eleven dimensions that he called M-theory. Witten's deductions sparked what became known as the second superstring revolution. It finally seemed likely that M-theory would prove to be the unified field theory that Einstein had struggled to discover during the last thirty years of his life: the theory that would unite general relativity and quantum mechanics. M-theory seemed to be the Dark Tower of physics—but the Crimson King had one last card to play.

The Secrets of the Universe
(a Simplified Version)

It's more than ten years past that second revolution, and so far, no one has been able to describe the underlying principle of M-theory—which has some scientists wondering whether superstring theory isn't really the final answer to the mysteries of the universe that its supporters believe. One such physicist is Dr. Lawrence Krauss of Case Western Reserve University in Cleveland, who called string theory "a colossal failure."[8]

According to Dr. Krauss, "We bemoan the fact that Einstein spent the last thirty years of his life on a fruitless quest, but we think it's fine if a thousand theorists spend thirty years of their prime on the same quest."[9]

Other critics of string theory are watching and waiting for new discoveries before rendering opinions. The biggest problem of string theory is the lack of any experimental evidence for strings or even a single experimental prediction that could prove the theory correct.

Strings are thought to be so small that string effects would show up only when atomic particles are smashed together at huge energies, approximately 1020 billion electron volts. Unfortunately, that would require a particle accelerator bigger than anything that could be constructed on Earth, leaving such tests entirely in limbo.

The huge gap between theoretical speculation and testable reality has led some critics to suggest that string theory is more philosophy than science. Other critics feel that the theory is still too vague and that some promising ideas have yet to be proved rigorously.

Somewhat less pessimistic is the Italian physicist Dr. Daniele Amati, who once joked that string theory was a piece of twenty-first-century physics that had fallen by accident into the twentieth century—and that it would require twenty-second-century mathematics to solve.[10]

In summer 2004, in Aspen, Colorado, where Dr. Schwarz and Dr. Green presented their first paper on string theory, the two scientists and a number of other physicists celebrated the twentieth anniversary of the event. Still, even there, the biggest names in string theory had to admit that after twenty years, they still did not know how to test string theory. Nor were they sure exactly what the theory meant. Meanwhile, their quest for answers using M-theory was proving to be stranger than they had ever conceived.

In M-theory, the universe has eleven dimensions—ten dimensions of space and one of time. Plus, it also contains extended membranes of various dimensions, known generically as "branes."

Using this new model of the universe has stunned the minds of many cosmologists. Some theorists now believe that our own particular universe may be a four-dimensional brane floating in some higher-dimensional space, resembling an air bubble in a fish tank, perhaps with other branes—parallel universes—drifting nearby. One fascinating theory proposes that a collision between two branes might have been responsible for the big bang. Or perhaps some other interaction between branes produced the dark energy in our universe. No one is sure about the answers, but the questions are fascinating.

One of the greatest successes of string theory involves the study of black holes. According to Einstein's theory of general relativity, black holes are bottomless pits in space-time, devouring everything, including light, that gets too close. In string theory, black holes are a dense jungle of strings and membranes.

In 1995, Dr. Strominger and Dr. Cumrun Vafa, both of Harvard, were able to perform a huge number of string theory equations that calculated the information content of a black hole. Their answers matched a famous result obtained by Dr. Stephen Hawking of Cambridge University, who used more indirect means to obtain his solution in 1973. These calculations offered new proof that string theory works for the real universe.

Another equally important string theory discovery was that the shape of space is not fixed but can change. Even more interesting, string theory proved that space can also rip and tear. This information is of particular interest to scientists concerned with traveling to other stars. As shown on numerous TV shows, the best way to travel long distances in space is by ripping a hole in the fabric of the universe, traveling through the fourth dimension, and then tearing another hole to get out—the stuff of science fiction for many years, but now proved possible by M-theory. If we ever leave our solar system for other stars by traveling through hyperspace, it will be because of M-theory.

Scientists hoped to have some measure of tangible proof of superstring theory in 2007, when the Large Hadron Collider was expected to be turned on at CERN in Geneva, Switzerland. Although the schedule for turning on the collider has been delayed slightly, when the process is completed the collider will slam protons with seven trillion volts of energy apiece. It's possible, according to one version of superstring theory, that such collisions could create black holes or particles disappearing into the hidden dimensions. Another possibility of what the collider might find is supersymmetry. Supersymmetry states that a whole group of ghostly elementary particles symmetrical to photons, quarks, and electrons should exist. Scientists believe that the lightest of these particles would have a mass-energy that could be detected by the collider. Even Dr. Krauss admits that the existence of supersymmetry would serve as a strong indicator that superstring theory is true.

At present, superstring theory and M-theory are still considered by many physicists to be the best hope for finding Einstein's dream, a unified field theory that will amount to a theory of everything. That day is still far in the future. Still, according to a talk given by Dr. Witten in October 2004, "It's plausible that we will someday understand string theory."[11]

Like Roland, we hope we will defeat the Crimson King known as ignorance and climb up the steps leading through the eleven dimensions of the Dark Tower to the top. Once there, armed with the knowledge of how the universe works, we suspect that like Roland, we will once again find ourselves on the first leg of another long journey of discovery.

6

TRAVELING IN TIME WITH STEPHEN KING

The Langoliers

This is the past. It is empty; it is silent. It is a world—
perhaps a universe—with all the sense and
meaning of a discarded paint-can.

—*The Langoliers*

W hile time-travel stories are common, nobody writes them with the bizarre twists of Stephen King novels. One case in point is *The Langoliers*, where the horrors of airplane travel take on new meaning.

Time Was

Stephen King's *The Langoliers* is one of a number of short horror novels that recycles the plot of an old 1930s melodrama in a slightly new and inventive fashion. A group of disparate people, young and old, rich and poor, are trapped together by circumstances in a deserted location. For one reason or another, there's no possibility of rescue from the outside world. Hunting them is an implacable monster (or monsters) that wants to eat, ravage, or destroy them all. Plus, there are major personality conflicts in the group itself, which prevents them from cooperating in a manner that would be safest for all. It's a scenario that works for Agatha Christie's *Ten Little Indians* right up to King's *The Stand*, with only minor variations in the plot. Though the basic story line is familiar, this doesn't mean that in the hands of a talented author it can't be done in new and different ways. The very familiar made terribly unfamiliar is one of the best methods of generating fear and suspense in a novel. It works quite well in *The Langoliers*.

The story is set in modern times and begins at night at Los Angeles Airport. Brian Engle, a pilot for American Pride airlines, has just landed a plane from Tokyo, having battled a pressurization problem for most of the trip over the ocean. He's ready for a shower and a good night's sleep. That's when an executive of the airline tells him that his ex-wife just died in a fire in Boston. Though he's exhausted and dead tired, Brian accepts the offer to take the redeye to Boston so he won't miss the funeral. As soon as he gets comfortable in his seat on the plane, he falls asleep.

While he sleeps, other passengers get on board and make ready for takeoff. Dinah Bellman is a little girl, blind since birth, who is flying with her aunt to visit an eye surgeon in Boston who thinks he can restore her sight. Albert Kaussner is a Jewish teenager, a violin prodigy, who is scheduled to play in a concert. And there are others; the plane is half-filled with people when it takes off.

Approximately an hour after takeoff, however, Dinah wakes up from a particularly bad dream and panics when she discovers that her aunt is no longer seated next to her. Dinah's screams waken nine other passengers. They're the only ones left on the plane, which is still flying east. Everyone else, including the pilot and the copilot, has vanished. The mystery only deepens when the passengers search the seats around them and find earrings, wallets, wigs, and other intimate and personal items that must have belonged to the missing passengers. Worse, Albert finds a pacemaker that obviously came from a living body and metal joints that were in a man's knee. It's difficult to understand, but the ten survivors are forced to accept the fact that everyone who was awake on the plane has melted away into nothingness. Only their small group—the people who were sleeping—is left.

Brian takes charge of the survivors with the help of Nick, an Englishman who seems quite capable of maintaining order. The two men break into the cockpit to find that, as they suspected, the entire flight crew is missing. When Brian tries to report their trouble to the ground via the plane's radio, no one is on the airwaves, not even the government advance warning networks. Brian gets spooked, especially when he notices that the city of Denver below is entirely without lights. The possibility of atomic war flits through his thoughts, but no scenario that any of the remaining passengers can come up with makes any sense. Brian diverts the plane from its route to Boston, opting to land it at the airport in Bangor, Maine, which has a much longer runway.

One extremely belligerent passenger, Craig Toomy, can't seem to comprehend the horrifying mystery that's engulfed the plane. He's entirely self-absorbed in his own problems and tunes everyone else out. He has to make it to Boston by the next morning for a business conference, and he screams and screams, demanding that the plane stay on its original route. His threats are annoying but make no sense. After a few minutes, he is routinely ignored by the rest of the passengers.

Also on the flight is Laurel, a schoolteacher in search of a perfect man; Betheny, a rebellious teenage girl; and Bob Jenkins, an elderly mystery novelist. The other passengers play much smaller, incidental roles in the story. It becomes clear to everyone fairly quickly that they survived whatever happened to the rest of the people on the plane because they were asleep when the disaster occurred.

Brian is able to fly the passenger jet to the Bangor airport, but, much to everyone's disappointment, they find the place deserted. Worse, the air seems stale and dead. There are no odors in the air, and all sound is muted. There are no echoes. The food in the terminal snack bar tastes like straw.

It's then that Bob Jenkins finally comes up with a theory about what has happened. They've been wasting their time trying to learn what happened to everyone else. Instead, they should be worrying about what happened to them because he feels certain that the world has not changed.

Instead, he proposes that the plane flew through "a hole in the time stream," that reality itself had a rip in it and the airplane plunged through it and went fifteen minutes into the past. According to Jenkins, "This is the past. It is empty, it is silent. . . . The world is clearly unwinding around us. Sensory input is disappearing. Electricity has already disappeared."[1]

When Albert asks Jenkins whether they might not have flown into the future instead of the past, the mystery writers admits that he doesn't know for sure but he's pretty convinced that they are in the past. The future would be just starting. That's when Dinah speaks and puts into words what everyone senses: "It feels over."[2]

The passengers hear strange noises in the distance, noises that seem to be drawing closer. No one is sure what the sounds mean, other than Craig Toomy, who says they are the noises made by the Langoliers, a childhood monster used by his father to frighten him when he got behind on his schoolwork. The Langoliers are creatures that eat children who waste time. Though the passengers believe that Toomy is crazy, his explanation of the noises sounds ominous. The sounds are not the noises of anything recognizable; they are unnatural, horrifying.

The problem the passengers face is that none of them can figure out a way to return to the present. They split up into groups, trying desperately to find some answers in the deserted terminal. But Craig Toomy isn't concerned with answers. Toomy has gone over the edge. He's become convinced that Dinah, the little blind girl, is the cause of all of his problems. So when he finally gets the chance, he stabs her in the chest with a kitchen cleaver. Miraculously, Dinah doesn't die, although she is very badly injured. Toomy manages to escape, and he hides somewhere in the terminal. Nick wants to hunt the man down and kill him, but Dinah tells him that Toomy shouldn't be killed. She doesn't explain why, but she convinces Nick that the madman must be allowed to live.

Finally, Brian realizes that the one chance to return to the present is to fly through the same rift in time they went through on their trip east. If the plane flies through the rift going west, hopefully it will emerge in the present. Dinah, who is near death, is loaded into the plane, and the rest of the passengers search for fuel for the trip.

Toomy kills another passenger but is horribly injured by Albert in return. Nick has a chance to kill the madman but, remembering what Dinah said, leaves him alive.

Just as the last of the fuel is being pumped into the airplane, the Langoliers arrive at the airport terminal. They are horrifying monsters that are devouring everything in sight, including the land, the buildings, the machinery, anything real. They closely resemble the monsters in the video game Pac-man, although they're many times larger—round creatures with huge white teeth. In their wake, they leave only emptiness. As Brian nudges the plane into take-off mode, Toomy stumbles onto the runway. The madman weaves around on the pavement and distracts the Langoliers, which follow him instead of the departing airplane. If not for Toomy's actions, the plane and all the passengers would have been devoured.

Bob Jenkins puts into words what the Langoliers are: "Now we know . . . what happens to today when it becomes yesterday, what happens to the present when it becomes the past. It waits—dead and empty and deserted. It waits for them. It waits for the time-keepers of eternity, always running along behind, cleaning up the mess in the most efficient way possible . . . by eating it."[3]

After their escape, the passengers realize that there is one final problem to be solved if they want to return to the present. They need to be asleep when they pass through the rift in reality. However, someone has to be awake at the controls to keep the flight steady. Nick volunteers to fly the plane althrough the hole in time, although he knows he will die doing it. Nick vanishes just as he guides the flight through the rift to the other side.

The plane comes in for a landing at LAX, but the world still appears to be empty of life. The time rip, working in the opposite manner, has sent them fifteen minutes into the future. When time catches up with them, the world once again fills with people and sounds and smells. The passengers have made it back to the normal world.

The Langoliers takes several of the basic premises of modern science fiction and develops these ideas to their full potential. Time is not a river in the King novel but a wave rolling forward and never turning back. Once time has passed a moment, it never returns to that moment again. Time travel in the usual manner is impossible, because the future doesn't yet exist and the past has been destroyed. The only time in the story is the present—because that's the only time that actually exists.

The Langoliers is one of the most original time-travel stories ever written. It takes a new, different, and unique look at a theme most readers felt was long used up. It's one of King's most imaginative works and deserves more attention than it has received. But could it ever happen? Is time travel possible? What about a rip in space-time? These are questions we'll address in the rest of this chapter, but first we need to know a little more about time travel and how it is supposed to work—at least, how writers before King thought it worked. With those ideas firmly in place, we'll be able to analyze the events in *The Langoliers* and make sure that they are free of the greatest enemy of time-travel stories: paradox.

A Short History of Time Travel

To fully understand *The Langoliers* and what makes its main concept so unique, we first need to examine the notion of time travel. Because there are no actual historical examples of time travel in the real world for us to study, we'll first look at time travel in fiction. This brief examination will help us define the risks and dangers associated with the concept. Plus, the survey will tell us where the concepts of *The Langoliers* fit in with other ideas on the subject. Then, because this book is about science, we'll make an attempt to understand the science of time travel.

Travel to the Past

The first time-travel story, as best as anyone can tell, was "The Clock That Went Backward," by Edward Page Mitchell, an editor with the *New York Sun* (reprinted in *The Crystal Man*, edited by Sam Moskowitz, 1973). Published in the *Sun* in 1881, the story told of two boys who find a broken clock, which, when it is wound backward, transports them to sixteenth-century Holland. The story was remarkable in part for being so unremarkable. It was actually pretty dull.

The first full-length novel that featured a man who traveled back in time was Mark Twain's *A Connecticut Yankee in King Arthur's Court* (1889). In the Twain story, the hero, Hank Morgan, a Connecticut factory superintendent, is knocked unconscious by a crowbar during a fight with one of his employees. When he wakens, he finds that somehow he's traveled back in time to the days of King Arthur. Using advance knowledge of a solar eclipse, he convinces the king's court that he is a sorcerer and is given power over the kingdom. Hank proceeds to build all the modern conveniences of the nineteenth century in seventh-century England. He also helps Arthur to modernize barbaric British society. In doing so, Hank makes enemies of Merlin the magician and the medieval church. After a number of satirical adventures, all of Hank's wonderful inventions and social reforms are destroyed by his enemies, and Hank is put into a magical slumber by Merlin for 1,300 years. He awakens in the present, gives a copy of his memoirs to Mark Twain, and then dies.

Exactly fifty years later, the science-fiction author L. Sprague de Camp wrote a novel, *Lest Darkness Fall*, that was a modern take on the Twain story. In the de Camp novel, his hero, a history professor named Martin Padway, is hit by a lightning bolt, which sends him back to the last days of the Roman Empire, at the beginning of the Dark Ages. Faced with the knowledge that hundreds of years of barbarism are about to commence, Martin decides to use all of his knowledge of modern technology to make sure that the Dark Ages never engulfs Europe. The de Camp novel is an entertaining riff on *A Connecticut Yankee*. In the Twain novel, history remains unchanged, but in the de Camp novel, history is completely rewritten. At the end of the book, the barbarians are routed and darkness does not fall.

The de Camp novel remains a classic of science fiction and is viewed by many science-fiction fans and scholars as the first "alternate history" novel where a man from the present travels backward in time and successfully changes events in the past. A year later, in 1940, de Camp wrote a short novel, *The Wheels of If*, where the hero finds himself thrust into one parallel world after another, each the result of history having taken a different turn centuries earlier.

Travel to the Future

The first time-travel story to deal with travel into the future instead of the past was H. G. Wells's famous novel *The Time Machine*. The story was published in book form in 1895. In Wells's novel, a nameless inventor builds a time machine and uses it to visit the world of A.D. 802,701. In the far future, the hero discovers that mankind has branched into two races, the gentle Eloi and the brutish Morlocks. The two species of humanity live in a garish sort of symbiosis, with the Morlocks keeping the world running smoothly for the Eloi, while the Eloi serve as the Morlocks' food. After a grim adventure rescuing his time machine from the Morlocks, the time traveler advances even further into the future to witness the dying days of Earth. Wells wrote his novel as a social commentary on English society at the end of the nineteenth century, but over the years, his message was forgotten and his book slipped into the realm of popular literature.

Time Paradoxes

Written a few years before Wells's novel, F. Anstey's humorous fantasy *Tourmalin's Time Cheques* was the first novel to discuss the paradoxes that might be caused by manipulating time. Anstey was a British humor writer, many of whose novels routinely used themes that later became standards in the fantasy field. His *The Brass Bottle* was filmed in 1964, with Barbara Eden, and was an uncredited but fairly obvious inspiration for the TV show *I Dream of Jeannie* a year later.

During the 1930s and 1940s, time travel emerged as one of the most popular themes in the science-fiction and fantasy pulp magazines. The pulps printed the earliest works of Isaac Asimov, Robert A. Heinlein, John D. MacDonald, Raymond Chandler, and many other writers. The first few time-travel stories published in the pulps were mostly variations of the "grandfather paradox." In such stories, the lead character goes back in time and intentionally or unintentionally kills his grandfather or some even more distant ancestor. With his older relation dead, the protagonist is never born. Since he never existed, he is therefore unable to return to the past, meaning the killing never took place, and he is born, returns, is killed, and on and on.

The grandfather paradox has no straightforward logical solution, which made it a challenge for science-fiction writers of the period to play with its premise. Typical of the stories written about the paradox is "I Killed Hitler" by Ralph Milne Farley (*Weird Tales*, July 1941), in which the protagonist does exactly what the story title proclaims, but ends up taking the dictator's place and becoming Hitler.

Time Is a River

The time-travel stories of the 1930s and 1940s were based on the concept that time was a river with a current moving at a set speed. Humanity was the passenger in a boat, and the scenery on the shore consisted of modern-day events. Taking that analogy one step further, time travelers were people who, by one means or another, climbed out of the boat and jumped down the river into the past or swam up the river into the future. The results of their actions made up the stories.

Fortunately for science-fiction fans, John W. Campbell Jr., the editor of *Astounding Stories*, the leading science-fiction magazine of the period, insisted on strong logical content in all the fiction he published. Campbell's strong editorial hand resulted in some of the most intelligent time-travel stories ever written. Campbell also served as the editor of *Astounding*'s companion magazine, *Unknown*, which first printed *Lest Darkness Fall* in 1939.

Time Loops

In Robert A. Heinlein's "By His Bootstraps" (*Astounding SF*, October 1941), a man is kidnapped into the future and spends the rest of the story going back and forth in time searching for his kidnapper, only to realize in the end that the person who initiated the crime is himself years older. In "As Never Was," by P. Schuyler Miller (*Astounding SF*, January 1944), a time traveler takes a trip to the future and wanders into a museum where a strange, unearthly artifact is on display. Fascinated by the weird device, the time traveler steals the object from its case and then returns to the present. Scientists study the item but can't explain what it is. After years and years of experiments with no results, the piece ends up being placed in a museum, where it will someday be stolen back to the past. The hero has created a time loop, and the mysterious artifact has neither beginning nor end.

Equally diabolical is Henry Kuttner's 1945 short story "Line to Tomorrow" (*Astounding SF*, November 1945), where an ordinary man discovers that he can eavesdrop on phone calls made in the future. He listens, only to hear the two people on the line discussing the mystery of why he went mad. Knowing that he is going to go insane and there is nothing he can do to change the future, the protagonist slowly does go crazy.

Kuttner's wife, Catherine Moore, did him one better in "Vintage Season" (*Astounding SF*, September 1946), made into a TV movie in 1992 called *Disaster in Time*. An innkeeper rents out his building to a group of mysterious visitors, who, after some deducing on his part, he realizes have come from the future. It's not until the close of the story that he finally grasps the fact that they have come back to witness firsthand a terrible disaster. But again, there's nothing he can do about it when a deadly plague breaks out.

Time as a Chimera

One of the most interesting time-travel stories to appear in the 1940s was "Zero A.D." by "Lee Francis" (*Fantastic Adventures*, February 1948). Francis was a house name for the Ziff-Davis publishing chain, so the true author of the short novel will never be known.

In the story, the newspaper reporter Johnny Sharp is sent to interview a scientist at a nearby college for a Sunday magazine feature on science in history. Professor Crockett tells Johnny that people are unhappy even in the best of times because they sense that some great truth is hidden in the past. The professor feels that too often people look to the past for guidance instead of thinking about the future. Crockett has invented a machine, the Memory Finder, which allows people to accurately recall all of their memories back to their earliest childhood.

Convinced that the professor is on the track of a big discovery, Johnny helps him with his research and even gets his girlfriend to volunteer to try the Memory Finder. Strangely enough, none of the test subjects can remember anything before the year 1925. Finally, when one person does remember, he recalls his life on another world, Moneta, a place where scientists were planning some sort of huge experiment involving violence. The professor and Johnny realize that Earth is merely a gigantic test ground, populated by hundreds of millions of brainwashed citizens from another existence, with all evidence of the past manufactured to fool the population into thinking that man has always lived on Earth.

Modifying the basic design of the Memory Finder, the professor is able to project his mind and Johnny's to Moneta, to see whether they can learn more about the gigantic experiment. Moneta is a utopia, but the people there are no happier than those on Earth. Using a Memory Finder on a citizen of Moneta, the professor is shocked to discover that the man can remember back only twenty-five years. Moneta, like Earth, is also an experimental world with no real history. The story ends with Johnny and the professor wondering whether any reality has a past.

"Zero A.D." is hindered by poor characterization, subplots that made little sense, and unbelievable dialogue; however, it has a plot filled with surprises and a very different take on time travel. It deserved more attention than it received, mostly because it was ahead of its time.

A much more sophisticated telling of a similar plot is the basis for the novel *Strata*, by Terry Prachett (St. Martins Press, 1988). The novel begins with technicians busily planting dinosaur bones on planets that are slated for colonization. They need to bury evidence of evolution on the new world so that settlers (who have their memories wiped clean before landing) will be fooled into thinking that the human race evolved on that world. The whole situation turns bizarre when the technicians discover a pancake-flat world drifting along in interstellar space. They send three of their own to investigate this mystery world and discover that the galaxy is a much stranger place than any of them ever suspected.

Tampering with Time

By the 1950s, intelligently written science fiction had spread to more than a dozen different magazines, but time-travel stories had settled into a fairly predictable pattern. People from the future or the present meddle with time travel and bring about a disaster. Or they find themselves involved in a paradox that traps them in an unending time loop. Ray Bradbury's "A Sound of Thunder" (*Planet Stories*, January 1954), which we will mention in chapter 7, is a perfect example. The story deals with a hunting expedition that travels back to the time of the dinosaurs. A hunter panics and kills a butterfly. When he returns to the present, he discovers that he now lives in a fascist dictatorship. In "Beep," by James Blish (*Galaxy SF*, February 1954), police in the future are warned of impending crimes by radio signals from even further ahead in time. The one law of the police force is, don't ever mention the death of an officer on the air.

Ingenious Plots

Perhaps the wildest time-travel story ever written regarding past-present-future paradoxes is Robert A. Heinlein's "All You Zombies" (*Fantasy & SF*, March 1959). Through a series of misadventures with a time machine and incredible technology, the hero of the story ends up being not only his father but his mother and all the other characters he encounters in the adventure.

In the 1970s, time-travel adventures gave way to alternate world/parallel world adventures. See chapter 7 for our discussion of parallel worlds and the science behind them. Poul Anderson's

"Time Patrol," and H. Beam Piper's "Paratime Police" feature organizations of dedicated cross-dimensional cops who patrol the many branches of the time stream, making sure history isn't changed by meddling interlopers from the future. Meanwhile, Harry Harrison's *The Technicolor Time Machine* (1967) deals with the problems of filming an epic involving the Vikings on location in the past, while David Gerrold tried to include nearly every time paradox ever discussed in his short novel *The Man Who Folded Himself* (1973). Working in the opposite direction, Jack Finney's novel *Time and Again* (1970) demonstrates how altering events in the past by the slightest degree, if done at precisely the right moment, can cause tremendous changes to the future.

Collision *Course* (1973), by the British author Barrington J. Bailey, is one of the most inventive time-travel novels ever written. It takes the notion of time being a river and turns it inside out. In the story, soldiers for the militaristic future government of Earth discover mysterious ancient ruins across the globe where years before no such ruins existed. It soon becomes clear that the ruins are somehow growing younger. A civilization in the far future is on a time track extending backward from the future to the present. That time track is on a collision course with modern society, which goes from the present to the future. Unless something is done, the two time-crossed worlds are destined to collide in an explosion that could destroy all of time. In this situation, the river doesn't just flow forward; one branch flows backward. More than three decades after publication, *Collision Course* remains one of the most startling time-travel novels ever written.

Time Is Not a River

These stories are but a sampling of the hundreds of time-travel stories that were published in the century following *A Connecticut Yankee in King Arthur's Court*. Nearly every one of them relies on the concept that time is a river and that travel into the past or the future is merely moving somehow up or down the river to the particular time period desired. All of these stories emphasize that one person can change the past and thus, by his or her actions, change the future. Only one story treats time travel in a totally different manner.

That story, perhaps the most original time-travel story written before the publication of *The Langoliers* is "Yesterday Was Monday" by Theodore Sturgeon (*Unknown*, June 1941).

In the story, Harry Wright goes to sleep on Monday night and wakes up on Wednesday. The only problem is that Tuesday never happened. Harry never sleeps more than six hours, and he remembers falling asleep Monday night. But he feels sure that it's Wednesday, which makes no sense.

What makes even less sense is when Harry discovers little three-feet-tall men placing dust on top of windowsills, knocking dents into car fenders, and putting holes in wood steps. Worse, Harry appears to be the only normal-size person around, and all of the little men keep talking about getting the scenery ready for "the actors."

Only after Harry meets Iridel, a supervisor from Thursday, does he learn that all of reality is a huge play and that the human race serves as the actors. The world is scenery, which needs to be changed and updated for every act, and the little men are stagehands. Harry is an actor who has somehow stumbled backstage.

Time has no meaning, according to the supervisor. He compares time to a road of blocks. Each day the actors move onto a new block and the play continues. Ahead of them, the stagehands build the next block. Behind them, the stagehands remove the old blocks that have been used, sending the pieces of scenery that aren't too badly damaged further into the week.

Sturgeon played the story for laughs, but the idea behind the story is actually terribly disquieting. Shakespeare wrote that "all the world's a stage," but Sturgeon was the first writer to use that line as the basis for a story. "Yesterday Was Monday" straddles the line between fantasy and science-fiction in that the producer shares a lot in common with God but is producing the play for an unnamed audience. And it is strongly implied that death is merely a way of retiring an actor and converting him into a stagehand.

"You're Another," by Damon Knight (*Fantasy & SF*, June 1955), tells an entirely different story but also features the main character discovering that all of reality is nothing more than a play. It is an excellent story with a wonderful twist ending, but the Sturgeon story was first and is much more memorable.

"Yesterday Was Monday" is a time-travel story that takes two giant steps away from the concept of time as a river. Instead, it treats time as something that can be broken down into discrete intervals, a road of building blocks as described by Iridel. It is a very different way of treating time in fiction, so different that it remained unique for nearly fifty years. That's when, slightly more than a hundred years after Twain and approximately forty-nine years after Sturgeon, Stephen King used the concept to write a short horror novel—*The Langoliers*.

The Science of Black Holes

One reason *The Langoliers* is not listed as one of King's best works is that most people feel that it's too unbelievable. Readers are willing to suspend disbelief and accept such events as biological warfare wiping out 99 percent of humanity, dead animals rising from the grave, a girl with pyrokinetic powers, or cell phones emitting a tone that drives the users insane. Yet they're not able to accept the possibility that people can travel backward or forward in time.

Of all the science-fiction themes, time travel is considered the most outlandish. We know that spaceships to other planets are possible. Aliens living somewhere in the galaxy seems fairly reasonable to everyone who believes in flying saucers. Immortality, or at least people living for hundreds of years, is right around the corner. But there's no chance of time travel. Ask almost anyone and they'll tell you, time travel is impossible. Just make sure the person you ask isn't a physicist, because they know the truth. Time travel may be possible someday.

Physics and Time Travel

In chapter 5 we discussed Einstein's theory of special relativity, as well as Einstein's theory of general relativity. These theories have been proven true by observing hundreds of examples in our universe. Special relativity combines space and time into a single continuum that we call the space-time continuum. General relativity explains gravity as space-time warped by mass. Physicists use Einstein's equations to describe the universe by a system of "field equations." There's nothing in these equations that proves time travel is impossible. In our universe, if something is not impossible, then it's possible.

The existence of powerful gravity fields implied extremely convoluted space-times. In 1949, the mathematician Kurt Gödel was the first scientist to discover that under certain specific circumstances, the theory of general relativity actually predicted time travel. Scientists working with the field equations discovered solutions that allowed for "closed time-like curves," which implied that time travel into the past was possible.[4]

We must admit that there's no evidence that time travel has ever taken place and no one is exactly sure how it would be done. Still, there are some interesting theories.

The most popular plan for building a time machine requires the designer to create a wormhole. A wormhole is a rip in the fabric of the universe that connects two black holes or a black hole and a white hole (a celestial entity that repels matter). That connection is made through hyperspace, which we discussed in detail in chapter 5. The question we didn't go into there is how to create a wormhole. To do that, we first need to examine black holes.

What Are Black Holes?

Black holes have become one of the science touchstones of the late twentieth century, yet most people have no idea what they are. You might be surprised how long the concept has been around.

A black hole is a concentration of mass so huge that the force of gravity warps space to such a degree that nothing can escape it. In other words, the gravitational field surrounding the object is so powerful that the escape velocity near the object exceeds the speed of light. This implies that nothing, not even light, can escape its gravity. Since light cannot escape the object or even be reflected from it, the object is invisible to the rest of the universe. It is a "black" hole because no light can come from it.

The original idea of a cosmic object so massive that no light could escape from it was first proposed by John Michell, an Englishman, in a paper sent to the Royal Society in 1783. The concept of gravity and escape velocity had been established by Newton nearly a hundred years earlier. Michell calculated that an object five hundred times the radius of the sun, with the same density as the sun, would have an escape velocity on the surface of the speed of light. This would therefore render it invisible to observers.

In 1796, the French mathematician Pierre-Simon Laplace independently came up with the same idea. It was published in the first two editions of his book *Exposition du Systeme du Monde* but then dropped.

The Schwarzschild Radius

In 1905, Einstein's paper "On a Heuristic Viewpoint Concerning the Production and Transformation of Light" established that light was made of photons, which were affected by gravity. In 1915, Einstein developed the theory of general relativity, which explained the relationship between mass, space-time, and gravity. Several months afterward, Einstein developed the theory of gravity called general relativity. A few months later, the German scientist and astronomer Karl Schwarzschild, while serving on the Russian front in World War I, completed the first two exact solutions of the Einstein field equations. The first solution yielded the answer for the gravitational field of a point mass, showing that theoretically a black hole could exist. According to the solution, a gravitating object would collapse into a black hole if its radius was smaller than a characteristic distance that became known as the Schwarzschild radius. Below this radius, space-time was so strongly curved that any light ray emitted in any direction in this region would travel toward the center of the system. This solution also implies that anything inside the Schwarzschild radius, including all parts of the object itself, would collapse into the center. The center would form a gravitational singularity, a point in space of infinite density. That's a nice way of describing a black hole.

The Schwarzschild radius is proportional to the mass of an object. An object that is smaller than its Schwarzschild radius is called a black hole. The formula for the Schwarzschild radius for an object with an escape velocity of the speed of light is:

$$r_s = 2Gm/c^2$$

where

r_s is the Schwarzschild radius.

G is the gravitational constant, that is, 6.67×10^{-11} N meters^2/$s^2 \times$ kg.

m is the mass of the object.

c^2 is the speed of light squared, that is $(299{,}792{,}458$ meters/s$)^2$ $= 8.98755 \times 10^{16}$ (meters/s)^2.

Plugging in all the numbers and doing the calculations results in an equation:

$$r_s = m \times 1.48 \times 10^{-27} \text{ meters/kg with } r_s \text{ in meters and } m \text{ in kilograms.}$$

For an object with the mass of Earth, the Schwarzschild radius is a mere 9 millimeters—about the size of a marble.

Perhaps the only writer to take black holes seriously in fiction was Donald Wandrei (the author of "The Blinding Shadows," mentioned earlier), whose story "Infinity Zero" first appeared in *Astounding SF* in October 1936. Wandrei describes in surprisingly accurate terms what would happen if a microscopic black hole were somehow created on Earth.

Black holes were an interesting result of the theory of general relativity, but no one thought they actually existed until the late 1960s. That's when a growing amount of evidence fueled speculation that gigantic black holes existed at the center of galaxies, formed by the collapse of huge clusters of stars. Since that time, more and more evidence, along with a number of papers on the theoretical properties of black holes, convinced most scientists that these massive singularities in space-time actually exist. As Professor Eric Blackman of the University of Rochester said in an interview with the authors, "During the past few years, observations of the orbits of massive stars in the center of our galaxy have strengthened the case that there is a compact object in the center of the galaxy with a mass of about 3 million solar masses. This all but offers proof that there is a black hole in the center of our galaxy. Since our galaxy is a typical spiral galaxy, this suggests that other galaxies most likely also have black holes in their centers."

Wormholes: A Shortcut through Time?

All of which leads us back to wormholes. Albert Einstein first proposed wormholes in 1935. He cowrote a paper with Nathan Rosen in which they showed that general relativity allowed for what they called "bridges." They theorized that there could be places where space-time is folded that allowed a transfer of matter from one point to another in the universe.

The Einstein-Rosen bridges concept didn't generate much interest after its initial appearance in print, possibly because black holes at the time were considered improbable at best. The idea remained dormant until the 1980s when Carl Sagan signed a lucrative contract for a novel about man's first contact with an interstellar species, titled *Contact*. Sagan needed to have his heroine leave Earth and arrive at another solar system without spending decades in outer space. Looking for a scientifically plausible explanation for hyperspace travel, he asked the UCLA physicist Kip Thorne for help. Thorne came up with much of the modern framework for wormhole theory.

The name *wormhole* comes from the following analogy: Imagine that the universe is the skin of an apple. A worm is crawling over its surface. The distance from one side of the apple to the other, measured by the worm's path on the apple, is equal to half the circumference of the apple. If, however, the worm burrows a hole directly through the apple, the distance it needs to travel is much less.

In 1988, Kip Thorne and a number of his colleagues at UCLA came up with the description of a time machine based on both quantum mechanics and general relativity.

According to Thorne, a wormhole is created somehow with both ends near each other. One end of the wormhole remains in place while the other is accelerated to nearly the speed of light, perhaps with an advanced spaceship moving inside. The far end of the wormhole is then brought back to the point of origin. Because of the effect of time dilation due to traveling at near the speed of light, the accelerated end of the wormhole has now experienced less subjective passage of time than the stationary end. Any object that goes into the stationary end would come out of the other end in the past relative to the time when it enters. One significant limitation of such a time machine is that it is possible to go only as far back in time as the initial creation of the machine. In essence, it is more of a path through time than it is a device that itself moves through time.

According to current theories on wormholes, creating a wormhole of a size big enough for a person or a spacecraft to travel through, keeping it stable, and moving one end of it around would require more energy than is generated by the sun in several billion years. Also, constructing a wormhole to use as an entrance and an exit would require "exotic matter," a theoretical substance that has negative energy density, to keep the mouth of the wormhole open. Exotic matter has not been proved not to exist, but no one is sure that if it does exist, it could be used to construct wormhole entrances. Constructing wormholes to use for time travel appears to be far beyond any science we can imagine at present.

Unfortunately, a new study by Stephen Hsu and Roman Buniy of the University of Oregon, argues that this method of building a traversable wormhole may be fatally flawed. In a recent paper, the authors looked at a kind of wormhole in which the space-time "tube" shows only weak deviations from the laws of classical physics. These "semiclassical" wormholes are the most desirable type for time travel because they potentially allow travelers to predict where and when they would emerge. Calculations by the two Oregon researchers show that a wormhole combining exotic matter with semiclassical space-time would be fundamentally unstable.

"We aren't saying you can't build a wormhole," said Dr. Hsu "but the ones you would like to build—the predictable ones—those look like they will fall apart."[5]

No one is 100 percent positive whether we will ever be able to build wormholes for time travel. Despite all of the immense challenges one would face when traveling in time, it's not absolutely impossible—which assures us that the events in *The Langoliers* might be extremely improbable but not impossible.

Where Are the Visitors from the Future?

Assuming there is some other method to build a time machine that requires less energy than the wormhole method, we are faced with the fairly obvious question: why haven't we been visited by time travelers from the future? Stephen Hawking is one of many scientists who has raised this point and used it as proof that time travel is impossible.

Hawking's notion was first mentioned in a paper written in 1992, where he stated that the laws of physics prevent time travel on all but submicroscopic levels. The belief in this notion was known as the chronology protection conjecture.

Actually, the point was first raised in the science-fiction story "Barrier," written by Anthony Boucher (*Astounding SF*, September 1942). Boucher's answer for the question was that an energy barrier had been erected in the future by a tyrant to prevent time travelers from journeying from the distant future to the past. The more obvious solution to the question was one used in a number of stories published during the same period that asked the same thing about aliens from space. The stories conjectured that we had been visited by time travelers, but that they had come in disguise. The reason for the disguises was not that the time travelers were afraid of being caught by the police, but mostly because they were afraid of causing a paradox that would somehow wipe out their future.

The Novikov Self-Consistency Principle

The apparent lack of time-travel paradoxes in history is a tempting rationalization for arguing that time travel is impossible; however, the lack of paradoxes has another explanation that doesn't rule out time travel. That explanation is known as the Novikov self-consistency principle.

The principle was proposed by Dr. Igor D. Novikov in the

mid-1980s to solve the problem of paradoxes in time travel. In its simplest form, the Novikov principle asserts that if an event exists that would bring about a time paradox, then the probability of that event taking place is zero. In other words, the universe will not allow time paradoxes.

Seeking to prove this rather astonishing statement, Novikov resorted to a model that could be examined using advanced mathematics. He considered a billiard ball being hit into a wormhole in a certain direction so that it would travel back in time and hit itself, thus stopping it from entering the wormhole in the first place. Studying the problem, Novikov discovered that there were many trajectories that could result from the same initial conditions and did not end up in creating a paradox. From this result, Novikov reasoned that the probability of such consistent events was nonzero. Since that was true, he therefore concluded that the probability of inconsistent events was zero.

Applying his billiard ball solution to normal time travel, Novikov reasoned that whatever a time traveler might do in the past, he would never be able to create a paradox. Unfortunately, while Novikov's principle asserts that such events such as the grandfather paradox can't happen, it doesn't actually explain how those paradoxes are stopped. There have been several scientific theories and dozens and dozens of science-fiction novels and short stories written about the strange ways the universe prevents paradoxes from occurring, but none of them provide a reliable, all-encompassing solution. For that, we need to look to quantum mechanics and the many worlds interpretation of quantum theory as proposed by Hugh Everett III in 1957.

Everett's Many Worlds Theory

According to Everett's theory, whenever a number of possibilities exist in a quantum event, the universe splits into multiple worlds (this is, universes), one for each possibility. Each one of these worlds is real, and everything is identical except for the one different choice. From the moment that choice is made, the worlds develop independently of one another. Moreover, each world exists simultaneously with the others, though remaining unobservable by the others. The Everett interpretation of quantum mechanics renders paradoxes meaningless since whenever a time traveler does something contrary to history, another branch of reality is created. And the time traveler becomes part of that reality, ensuring that he can never return to his original future. Still, we are again forced to ask, where are the time travelers who changed our past? Why are they hiding?

Of course, if the time travelers experienced the events described in *The Langoliers*, no one could fault them for wanting to remain hidden. Along with their nightmarish adventure, they were the only six survivors of a major airplane disaster. Plus, the story hinted that numerous other planes had disappeared in the rift in time as well. If the time travelers were foolish enough to tell their story to the government, most likely they would all end up in an asylum. Worse, if the government believed them, they'd probably be shipped off to a secret laboratory somewhere to be probed, examined, and questioned endlessly by government scientists anxious to duplicate the space-time rift.

Still, that doesn't explain where all the other time travelers are. Maybe, just maybe, they're attending the Time Traveler Convention.

The Time Traveler Convention

The first ever Time Traveler Convention was held at the Massachusetts Institute of Technology (MIT) on May 7, 2005. The convention was held with the express purpose of contacting time travelers who were visiting the present from the future. The convention was organized by Amal Dorai, with assistance from current and former members of the MIT group Pi Tau Zeta. This convention was the most publicized event ever held for time travelers and included mention on the front page of the *New York Times*. Assuming that time travelers from the future could visit any time or location they chose, this event seemed like the perfect place for them to assemble.

The convention took place at 22:00 EDT on May 7, 2005, and

was held in the East Campus Courtyard and in Walker Memorial at MIT. The location was also listed as 42.360007 degrees north latitude, 71.087870 degrees west longitude. The convention was announced in advance, and more than three hundred people from the present attended. The convention included lectures on various aspects of time travel from three MIT professors: Erik Demaine, a MacArthur genius grant winner; Alan Guth, an Eddington Medal winner for theoretical astrophysics; and Edward Farhi, the winner of numerous MIT teaching awards. A De Lorean DMC-12, the car featured in the *Back to the Future* film trilogy, was also on display, close to the spot where the exact coordinates of the convention were listed. Whether time travelers from the future would have known the significance of the car was not clear.

All time travelers from the future were welcome to attend the event and did not have to observe any particular dress code. Obviously, to avoid fraudulent claims, the organizers requested that the guests carry some sort of proof that they had come from the future. A demonstration of technology far in advance of anything available in 2005 was one suggested method of verification. Another method was for the visitor to bring along a cure for AIDS, cancer, or some other incurable disease.

Unfortunately, no time travelers attended, though we suspect they aren't looking for publicity. After all, someone might expect them to explain exactly how time travel works.

7

PARALLEL WORLDS

"The Mist" • *From a Buick 8* • *The Dark Tower*
The Talisman

Something in the fog took John Lee.
And I heard him screaming!

—"The Mist"

Beyond Time and Space

Some of Stephen King's works, such as *From a Buick 8* (2002) and "The Mist" (in *Dark Forces*, 1980, and *Skeleton Crew*, 1985), rely on the notion of parallel worlds, which we briefly touched on in chapter 5. To start, what are parallel worlds? If parallel worlds exist, where are they? Is it possible to travel to parallel worlds, and if so, how might this be done? This chapter attempts to answer those questions.

"The Mist" appeared in *Dark Forces: New Stories of Suspense and Supernatural Horror* in 1980, as well as in the *Skeleton Crew* anthology. In the novella, and in the audio version (*The Mist in 3D Sound*, 1987), a Mist is unleashed, possibly from a parallel universe.

David Drayton, the narrator of the story, explains that the Mist came after a huge storm had destroyed most of Long Lake, Maine. Drayton, his neighbor Brent Norton, and his son, Billy, go into town seeking supplies after the storm. His wife, Steffy, remains at home. Drayton, Norton, and Billy become trapped in the Federal Foods Supermarket. As the Mist consumes the entire town, the three see the Mist roll around the grocery store, along with grotesque creatures that kill everyone outside the store. The Mist is most likely a government project to contact another reality that went terribly wrong. These monsters are unlike anything that lives on Earth and are hostile to all life that they encounter. After two days in the grocery store, most of the people trapped inside with the three lead characters have gone insane. Drayton and Billy escape, and they drive through the ruins of Maine toward Hartford, Connecticut. They're seeking other survivors. We assume that Steffy is dead. We know that Brent Norton is dead, because he leads a group of people into the Mist, and all that remains of them is a blood-sopped rope. As the story ends, the Mist continues to expand, growing large and larger, as it slowly engulfs everything in its path.

In King's *From a Buick 8*, Pennsylvania state troopers Ennis Rafferty and Curtis Wilcox answer a call from a gas station attendant and return to the police station with an old Buick Roadmaster. The car's owner has disappeared, and there's no information in the car to help track him down. So the troopers decide to stash the Buick in a shed behind the barracks until the owner resurfaces. Wilcox senses that something is really weird about the Buick, and soon afterward, Rafferty disappears. It's as if the Buick is acting as a pathway between worlds. From time to time, things emerge from the car's trunk that are definitely not of this Earth, and other times, often when nobody expects things to happen, the Buick pulls someone into its trunk and the person disappears, never to be seen again. The troopers decide to keep the Buick hidden forever. After all, there's something mighty strange about the car, and if it ever gets out, people could die.

In 2001, Wilcox dies in a car accident. Ned, his eighteen-year-old son, starts taking care of the police barracks as a way to retain the memory of his father. In his own way, Ned becomes part of the troop. Then one day Ned peeks into the shed and sees the hidden Buick.

He gets behind the wheel. When the police sergeant finds him, the boy is dead white, with a steady black gaze and robotic movements. Ned warns the sergeant to get out of the shed, and then the Buick starts humming, with a "pulse" that is "almost certainly a kind of telepathy."[1] Ned explains flatly that the Buick killed his father, and that he, Ned, intends to kill the Buick. The car is a creature from another place and time. With a butane match and a gas can, Ned intends to "light its damned transporter on fire." This way, the door will be shut, he says, to the other world forever.

Ned fires a gun repeatedly at the car, and the sergeant yanks open the car door. As the door opens, purple light streaks out of the car. The humming pulse becomes as loud and violent as the "precursor waves before the tsunami."[2] The inside of the Buick disappears and is replaced by purple light, and this light is the way into the parallel world. This being a Stephen King novel, things go from bad to worse before the story of the strange Buick is finally resolved.

In both "The Mist" and *From a Buick 8*, a bridge is formed between our world and another place, not an alien world hundreds of light-years away, but a parallel world, a planet that is nearly identical to Earth but is a quantum step away. These parallel worlds are often called alternate realities because they consist of our planet where just one or two changes occurred in history that made them different from our reality. Obviously, some realities are very different from our world, as in "The Mist" or *From a Buick 8*. In those worlds, the defining moment for Earth took place millions of years ago and resulted in totally different life-forms developing on the planet.

One easy way to understand the concept of parallel worlds is to think of reality as a gigantic tree. Our world is the trunk of the tree, and it towers up straight to the sky. At certain points in history, however, the tree trunk branches into several limbs. These are events that could have happened in several different ways. Each branch is one of those ways. Moreover, each branch continues growing and can branch itself into other directions farther along. The same is true of parallel worlds. Once they have branched off the main trunk, they can diverge into numerous alternate realities bearing little resemblance to our world.

Remember that parallel worlds are *duplicate* Earths, with a change or two in the past that expanded over hundreds or millions of years to make them very different. The similarities still make them near duplicates as far as gravity, atmosphere, scenery, and most other attributes. If Earth is a piece of white paper, then an alien planet is like a piece of cardboard. A parallel world is like a slightly fuzzy copy.

The worlds of *The Dark Tower* and *The Talisman* are almost exact duplicates of our world, but they possess a few major differences due to occurrences in history that made them into alternate versions of our world. A parallel world and an alternate Earth are almost exactly the same thing. A parallel world has changes that are the result of differences on a quantum level between our Earth and the parallel one. In an alternate Earth, the differences take place on the physical level, so it is a parallel world where the history of the planet has chaged due to a change in events.

Along with aliens from outer space, time travel, and people with ESP powers, parallel worlds are a staple of science fiction. It's not surprising that Stephen King decided to use parallel worlds as the basis for several of his most popular novels. Let's take a brief look at parallel worlds in science fiction and then examine the possibility of such parallel worlds in real life.

Worlds around the Corner

Parallel-universe fiction usually doesn't focus on the events that create a new reality, but rather on what happens after the new universes come into existence. For example, popular stories center on what the world would have been like if the South had won the Civil War or what the world would be like now if the American Revolution had failed and the United States was still a British colony. While not all parallel-world stories deal with wars, the concept is popular with military science-fiction writers because wars usually have key moments when a decision one way or another will have a profound effect on the future.[3] In Ward Moore's parallel-world novel *Bring the Jubilee* (1953), the main character travels back in time to visit the past and accidentally changes history, at the Battle of Gettysburg, and thus creates a world where the North doesn't win the Civil War. In Ray Bradbury's classic story "A Sound of Thunder," a man who time-travels back to the age of dinosaurs changes modern history by accidentally stepping on a butterfly millions of years in the past.

Changing any past event leads to the creation of alternate worlds, which lead to many parallel worlds. Time travel stories are concerned mainly with what happens to the time traveler. Alternate world stories are concerned with what happens afterward to the world.

While the basic concept of alternate history was developed in the nineteenth century, the first anthology of alternate history stories ever published appeared in England in 1931. Titled *If It Had Happened Otherwise: Lapses into Imaginary History* (1931), the collection was edited by the British historian Sir John Squire. The book consists of a group of fourteen essays on what the world would have been like if certain events in history had turned out differently. Squires recruited a number of scholars from Oxford and Cambridge as contributors, as well as important writers independent of any college. Topics range from "If the Moors in Spain Had Won," to Winston Churchill's clever "If Lee Had Not Won the Battle of Gettysburg," written as if by a historian in a world where the Confederacy has won the Civil War. He is pondering the notion of what might have happened if the North had won. Needless to say, the Churchill piece serves as a perfect setup for the Ward Moore novel published approximately twenty years later.

It was in the pulp magazines of the 1930s and 1940s where alternate reality stories blossomed. In Nat Schachner's novelette *Ancestral Voices* (*Astounding Stories*, December 1933), a man travels back to the time of ancient Rome and discovers that his ancestry is not as pure-blooded as he thought. Less than a year later, in Murray Leinster's "Sideways in Time" (*Astounding Stories*, June 1934), a cosmic disaster shuffles various alternate realities with modern-day history, creating a wild scenario for the survivors.

Jack Williamson's novel *The Legion of Time* (*Astounding Stories*, May–July 1938) is the first story to describe two possible alternate future realities sending warriors back to the present, each to guarantee that his world comes into existence. It is one of the first stories to promote the notion that the event that changes history might not be of historical importance but instead can be the simple act of a child finding a toy. The scenario of two different realities whose existence depends on one turning point in history became a popular science-fiction plot.

William Sell's "Other Tracks" (*Astounding Stories*, October 1938) was the first story to suggest that traveling from the modern day into the past would rewrite current history. Tom Garmot and his nephew, Charlie Thorne, travel from 1938 back to 1857 in hopes of building a better battery to power their home-made time machine. That doesn't happen, but the two time travelers instead find themselves rewriting history on each trip, as they try to straighten out a deal for postage stamps that goes terribly wrong.

A year later, L. Sprague de Camp's novel *Lest Darkness Fall*, (*Unknown*, December 1939) popularized the concept of a modern man returning to ancient times and changing reality by using his knowledge of the future. We discussed the de Camp novel in chapter 6. The concept of alternate realities sending agents back in time to change the past was first proposed in A. E. van Vogt's short novel *Recruiting Station* (*Astounding Stories*, March 1942) and further developed in "The Flight That Failed," by E. Mayne Hull (*Astounding Stories*, December 1942).

The lead character in "He Walked around the Horses" by H. Beam Piper (*Astounding SF*, July 1948) is the English government agent Benjamin Bathurst, who is on government business in Prussia. At a road stop in Perleburg to change the carriage horses for his coach, Bathurst walks around the horse and is never seen again. He just disappears. The Bathurst incident, which was based on a true mystery served as introduction to what was the first (and one of the most entertaining) series of stories dealing with a police agency that polices the billions of alternate realities in the universe of universes, often called the "multiverse." It is the job of the Paratime Agency to make sure advanced technology is never sold by an advanced civilization to a less civilized society.

In Clifford Simak's *Ring around the Sun* (1953), which we discussed in chapter 5 on the Dark Tower series, the hero ends up in an alternate Earth of thick forests in which humanity never developed. A few years later, in 1954, Poul Anderson took the whole concept of policing the multiverse a step further with his Time Patrol series. The Time Patrol exists mainly to make sure that time travelers from the future don't return to the past with hopes of changing the future. Their job is to prevent the time line from straying from its proper course, no matter what price they have to pay.

The Man in the High Castle by Philip K. Dick was a groundbreaking science-fiction novel published in 1963. The story takes place in the present, but a present where the Germans and the Japanese have won World War II. The novel paints an intense picture of what life would have been like in the United States twenty years after a Nazi takeover.

The 1980s and the 1990s saw a boom in science-fiction adventure novels dealing with alternate realities. Leading the charge was Harry Turtledove, a writer with a Ph.D. in history and boundless energy for rewriting American history. His most popular novel was perhaps *The Guns of the South* (1992), where the Confederate Army is given the technology to produce AK-47 rifles and ends up winning the Civil War. His numerous other alternate history novels include a series in which aliens invade Earth during World War II.

Recently, Philip Roth's *The Plot against America* (2004) describes an America where Franklin Delano Roosevelt is defeated in 1940 in his bid for a third term as president of the United States. Instead, Charles Lindbergh is elected, leading to increasing fascism in the United States.

Parallel-world novels are among the most popular stories published in modern science fiction. It's not surprising that Stephen King used the concept in a nontraditional manner by creating two quite believable and suspenseful horror stories.

The Science of Many Possible Worlds

Alternate realities are also a staple of much current scientific speculation. Scientists now believe that there may be an infinite number of parallel universes containing "space, time, and other forms of exotic matter."[4] Some of the parallel universes may include alternate versions of each one of us here on Earth. If you have a large nose, an alternate you sitting on a nearly exact duplicate of Earth may have a small nose. If you are a patient, kind person, your alternate self may be mean and unruly.

Now, before you think we're crazy to claim that there are alternate versions of you all over in parallel universes, consider that, according to *Scientific American*, the "simplest and most popular cosmological model today predicts that you have a twin in a galaxy about 10 to the 10^{28} meters from here."[5]

To make this sort of estimate, scientists use probability statistics. They assume that outer space is infinite and filled with matter in a near-uniform way. They then apply various calculations and conclude that aliens live on other planets, and somewhere out there in the universe are our twins, just slightly different.[6]

Today's astronomers can see objects that are approximately 4×10^{26} meters away, a distance known as the Hubble volume or, more simply, as our universe. Scientists suggest that each of the infinite parallel universes has its own Hubble volume. A sphere with a radius of 100 light-years is located approximately 10 to the 10^{92} meters from our universe. This distant sphere is the same as a sphere centered in our universe. If you see something in this universe, then someone else in the other universe sees the same thing—or so the theory goes. Everything in that parallel universe equals everything here in our universe. There are minor fluctuations, such as somebody having a big nose rather than a small nose, but for the most part, says the theory, all is parallel.

It's quite possible that creatures can come from the gateway to other universes, as suggested in "The Mist." We're not so sure about Buick Roadmasters just showing up on planet Earth and sucking humans into another universe—that's a really big stretch. Of course, in the novel, one of the lead characters suggests that the Buick isn't really a Buick but merely a chameleonlike device that takes on the appearance of a car so it will blend in with the scenery. In any case, if the theory holds that everything is parallel, then a car here is a car there, and vice versa.

Multiverses and Chaotic Eternal Inflation

If that were not enough to comprehend and believe on some level, the theories about parallel universes continue by telling us that each universe is part of a gigantic multiverse. The idea of the multiverse is "grounded in well-tested theories such as relativity and quantum mechanics,"[7] and, according to some scientists, the question is not whether the multiverse exists but, rather, how many levels exist within the multiverse. We've mentioned the notion of a multiverse before, in chapter 5.

Every parallel universe that contains a near-identical you is called a Level 1 multiverse. Taken together, all of the Level 1 multiverses are called a Level 2 multiverse. Something known as chaotic eternal inflation is behind the idea of the Level 2 multiverse. This refers to the idea that space is expanding, a theory that has been substantiated since 1998 by observations of supernovas and cosmic background radiation. In the theory of chaotic eternal inflation, the word *inflation* refers to the expansion or stretching of space. *Chaotic eternal* has to do with the bubbles that form when space stops stretching in some spots. Eventually, an infinite number of these bubbles are formed in random, chaotic ways, and all of the bubbles differ due to the breaking of symmetry. We'll discuss this notion in greater depth later in this chapter.

Did Colliding Branes Create the Big Bang?

In December 2004, scientists observed a dozen galaxies that appear as double images. They postulated that what they observed is evidence of superstrings—concentrated threads of energy spanning millions of light-years across the universe. The vibration of superstrings varies and is directly related to the fundamental particles of matter, strings, which make up everything in the universe. This theory has become famous as the "theory of everything," which we also discussed in chapter 5.[8]

According to this theory, our universe is a three-dimensional "brane" that moves through the ten-dimensional space-time of superstrings. When two of these branes collided, scientists postulate that it created the big bang.[9] During this time, superstrings were created, as well as Dirichlet, or D, branes, which are inside other branes and serve as bridges between the branes. Furthermore, the D branes have only one bridge into one dimension of our universe. Of course, all of this is theoretical speculation. Nothing has been proved, but it does explain how a Buick Roadmaster can move from one parallel universe into another and how creatures from the Mist can show up one day outside a grocery store.

In January 2005, Michio Kaku wrote that scientists are speculating that

> our universe may be compared to a bubble floating in a much larger "ocean," with new bubbles forming all the time. According to this theory, universes, like bubbles forming in boiling water, are in continual creation, floating in a much larger arena, the Nirvana of eleven-dimensional hyperspace. A growing number of physicists suggest that our universe did indeed spring forth from a fiery cataclysm, the big bang, but that it also coexists in an eternal ocean of other universes.[10]

Multiple Realities

In addition to theories that predict multiverses far away from us are theories that predict multiverses that are right in front of us. These are the types of multiverses through which cars and aliens might travel. This sort of multiverse theory is also known as the many worlds interpretation of quantum mechanics. It was first proposed in 1957 by Hugh Everett III, which we discussed earlier in this book. We might also describe this theory as the Level 3 quantum many worlds multiverse. (Just don't try to say it fast ten times.)

Basically, this theory states that random quantum fluctuations occur that cause our universe to branch into infinite multiple copies of itself. For every question you answer, another version of you could have answered it another way. For every hair on your chin, another version of you might have the same hair, only a micrometer away.

Even if we consider only the major events in history as moments when the universe branches into two new realities, the number is staggering. How many major events have occurred in human history? Thousands, tens of thousands, hundreds of thousands, millions? Plus, every time a new reality is created, events in its future will also result in parallel worlds. And this branching effect has been going on since the beginning of history. So the number of branches is in the billions of billions.[11]

As Ray Bradbury demonstrated so well in "A Sound of Thunder," even the smallest change in the Jurassic Age could have changed all of history that came afterward. Even the first volcanic eruption on the newly formed Earth might have affected the air we breathe today. Every event that has ever happened must be analyzed when determining how many parallel worlds might exist and what they're like. The number of parallel universes that exist in direct relationship to our own world is a function of the number of events that have taken place since the creation of Earth.[12]

The many worlds interpretation was proposed in 1957 by Everett when he was a Princeton graduate student. According to Everett's theory, whenever multiple possibilities exist in quantum events, the world splits into many worlds, one for each possibility. These worlds are all real and exist simultaneously with the first while remaining unobservable by any of the others. In other words, every event in quantum mechanics that has more than one possible solution yields a parallel world for each answer.

Everett's theory is not well liked by many physicists who prefer the Copenhagen interpretation of quantum mechanics, developed by Niels Bohr and Werner Heisenberg in the late 1920s (in Copenhagen). The Copenhagen interpretation says that in quantum mechanics, measurement outcomes are basically indeterministic, so there is never a choice of answers. It just can't be determined for sure. As we mentioned in chapter 5, Albert Einstein was a strong opponent of the Copenhagen interpretation.

While the Copenhagen interpretation of quantum mechanics is fascinating, we'll stick with the more interesting many worlds theory. Working with that concept, we find that the number of alternate universes for Earth, while incredibly large, is finite. Since Earth has not existed forever, if we had a gigantic computer and a lot of spare time for calculating, we could come up with the number of all possible quantum events that have taken place since Earth was formed. We could thus calculate every possible parallel universe created by those events. From there, we could track down every possible quantum event that took place in all these branch universes. Continuing outward, following every possible branch, counting quantum events at the speed of light, we could record billions upon billions of probability worlds that are linked to the first quantum action on the planet Earth. The number would be mind-boggling. Still, the sum of an immense but finite group of immense finite numbers is a finite number. So, although the voyage would be staggering, since the number of universes created by Earth over its billions of years of history is finite, we could travel from the first created parallel universe to the last. Science-fiction stories that discuss the details of traveling across millions upon millions of parallel universes include works by Poul Anderson, Andre Norton, H. Beam Piper, and Keith Laumer.

But is that immensely huge number the total of all parallel universes that exist? We've dealt with the many worlds theory as it relates only to Earth. Earth is just one planet, however—part of one solar system, part of one galaxy, part of one galaxy cluster, part of our universe. Since the entire universe contains atoms whose particles are subject to the laws of quantum mechanics, every atom in the universe is subject to the many worlds theory. While the life cycle of planets, stars, and even galaxies is finite, our current theory about the universe states that it began with a big bang billions of years ago and has been expanding ever since, creating new stars, new solar systems, and new galaxies as it does. We have an infinite number of atomic particles whose movements create parallel universes throughout the entire universe. In other words, since there are an infinite number of atomic particles, there are an infinite number of parallel universes.

Schrödinger's Cat

The many worlds interpretation of quantum mechanics also ties in closely to the Schrödinger equation. Erwin Schrödinger was an Austrian physicist who lived from 1887 to 1961, and he was a pioneer of quantum physics. He devised a famous, yet imaginary, experiment involving a cat.

Schrödinger noted that when an atom decays, there might be a 1 in 10 chance that it will decay in thirty minutes, a 9 in 10 chance that it will decay within one day, and so forth. Furthermore, if you're observing the atom at a particular moment, it's either decaying or it isn't decaying; at that moment, there is a 50-50 chance that the atom is decaying. The atom is in a "confused" state, not knowing whether to decay or not decay within that particular moment. Now, what if you're not watching the atom?

Schrödinger suggested that in an imaginary experiment, we might put a "confused" radioactive atom in a locked box with a living cat. The atom alone could not hurt the cat, but if the decay of the atom triggered a killing device, then if the atom decayed, the poor cat would be killed. The question then arises: when the atom is in the 50-50 state, is the cat dead or alive?

The Copenhagen interpretation mandates that nothing is real unless you look at it. So if you open the box and look at the cat, you'll know whether it is dead or alive. If, however, nobody opens the door and looks at the cat, then there is no way to know whether the cat is dead or alive during the moment when the atom is in its 50-50 confused state. You could say that the cat is dead and alive at the same time.

In terms of the many worlds interpretation, what Schrödinger's cat theory tells us is that all the parallel universes are real, even though we can't see them. The cat is both dead and alive at the same time, according to the many worlds interpretation. That's because in one world the cat is alive, and in another world the cat is dead. If you exist in this world, open the box, and see a live cat, then another you in another world is opening the box and seeing a dead cat.

Each time anything in the quantum realm has a choice—to decay or not decay, for example—the universe splits in this way. All of these parallel universes exist simultaneously.

The multiverse theory suggests that different laws of nature might exist on infinite worlds. It also suggests that life might exist—in fact, probably does exist—on many other worlds. Just as it's very possible that aliens exist on other planets, it's very possible that they can travel to us, and vice versa, through doorways across parallel universes. Such theories raise the eternal question: are the events of "The Mist" and *From a Buick 8* merely the imaginings of a gifted writer? Or are they possible, or even probable?

The Existence of Parallel Worlds

The many worlds interpretation of quantum mechanics hardly created a ripple in the scientific community when it appeared in a paper by Everett in 1957. For decades, Everett's paper was barely noticed. Over the years, however, interest in his notion of many worlds grew until Everett's idea became recognized as one of the major building blocks of modern quantum theory.

Science-fiction and fantasy writers, such as Stephen King, Ray Bradbury, and many others, have written stories about other worlds, parallel worlds, where duplicates of us, of our civilization, our world, and even our universe exist, where the only thing different from ours is perhaps a girl's hairstyle. Everett's theory said that for the smallest change in the orbit of an electron, an entire universe would change. Most important, because of that minor change in frequency of that electron, the parallel world created would be impossible to enter from another universe such as our own. As discussed in chapter 6 on time machines, traveling from one universe to another would require the energy of two black holes and vast amounts of "exotic matter," a substance not yet found in our universe.

Everett's theory slowly but surely replaced the Copenhagen theory in quantum mechanics; however, many scientists refused to accept the many worlds version of quantum theory. What troubled them the most was that there was no evidence proving that parallel worlds existed—especially since travel to such parallel worlds was impossible.

Guth's New Cosmological Principle

All that changed with a "spectacular realization" by the scientist Alan Guth in 1981. That was the way Guth described his discovery of perhaps the most startling concept in modern cosmology, the study of the beginnings of our universe. Guth's discovery, labeled *inflation*, was the first major revision to the big bang theory since it was first proposed by the Belgian priest Georges Lemaître in 1933. It stated that the universe went through a period of rapid, exponential expansion that was propelled by "a repulsive gravitational force generated by an exotic form of matter."[13] Guth's realization solved many of the major problems that had been raised about the big bang theory. Inflation was an idea that revolutionized the way scientists looked at the cosmos. That's because recent scientific cosmological data, obtained from satellite telescopes, was consistent with the predictions of inflation theory, thus offering physical evidence as to the possible truth of the concept. Not only had Guth discovered a new cosmological principle, he had discovered one that was consistent with observations of the actual universe. In the next twenty years, with the launch of a satellite measuring gravity waves, we will know for certain whether inflation is fact, a discovery that will change the way we look at the universe and ourselves.

What makes inflation so interesting to us is that the theory states that infinite parallel universes must exist. If we assume that inflation is true, we therefore can conclude that parallel worlds are also true—all of which will be revealed in less than twenty years.

In the meantime, as we wait, we are left wondering exactly what inflation says and what problems in cosmology it solves. Let's start with the questions and then see how Guth's revelation solved them all.

Three Questions about the Big Bang

In chapter 5, we discussed the four fundamental forces of the universe—electromagnetism, gravity, the weak nuclear force, and the strong nuclear force. We noted that each of these forces is entirely different. Each has different properties and different strengths, with gravity being the weakest, and the strong nuclear force being the strongest. The question we didn't ask during our discussion of these forces was: why is the universe controlled by four different forces—especially when these four different forces are so different?

The standard model of quantum mechanics, also mentioned in chapter 5, united three of the four forces. Only gravity was not included in the mix (and that led to the discovery of string theory). Still, many physicists were unhappy with the contrived methods that connected the three forces in the standard model. They wanted to find a single, simple explanation that for the relationship among the three forces, a grand unified theory.

It was this grand unified theory that introduced a new idea to cosmology known as spontaneous breaking. According to the grand unified theory, at the instant of the big bang, all four fundamental forces of the universe were united in one force, dubbed the "super-force." As part of this superforce, all four forces had the same strength. This superforce existed in a state known as a false vacuum. A false vacuum was a vacuum state that was not at the lowest possible energy state. The instant the big bang took place, the superforce began to "crack." "Spontaneous breaking" took place and the four forces broke away from the superforce, one after another. Each force went from the false vacuum to the true vacuum, a vacuum at the lowest possible energy state.

The grand unification theory thus tied the three forces together into one; however, it raised a number of major questions that couldn't be answered. One such question was known as the mono-pole problem. A monopole is a single magnetic north or south pole. In our universe, these poles are always found in pairs. Yet the grand unified theory predicted that a large number of mono-poles existed at the beginning of time and thus should still exist. Nonetheless, scientists searched the universe for centuries and never found a monopole.

Another problem raised by the grand unified theory was known as the flatness problem. The big bang theory couldn't explain why the universe as observed by us appeared so flat. If the universe exploded in a huge bang, it was logical to expect space-time to be curved. Mathematics applied to the problem predicted that the universe was curved, yet it seemed flat.

A third problem with the big bang, as described by the grand unified theory, was the horizon problem. To an observer, the night sky seemed relatively the same, no matter which direction a person looked. Telescopes confirmed that the universe was fairly uniform. Moreover, space satellites registered cosmic microwave radiation as uniform from all directions in space. And the temperature of background radiation in space deviated less than a thousandth of a degree wherever measured.[14]

These measurements were the basis of the horizon problem. The speed of light was the fastest speed allowed in the universe. Yet microwave radiation from one direction in the universe was nearly the same temperature as microwave radiation from the opposite direction in the universe. These two points were located 26 billion light-years from each other (each point being approximately 13 billion light-years from us). As the cosmologist Michio Kaku put it, "Since they are at the same temperature, they must have been in thermal contact at the beginning of time. But there is no way that information could have traveled from opposite points in the night sky since the Big Bang."[15]

Taken together, these three problems made cosmologists miserable. Their grand unified theory didn't seem very grand with three major questions about it unanswered. Then along came Guth with a startling concept that not only solved all three major problems but also revolutionized cosmology. As mentioned earlier, Guth called his solution inflation.

Guth proposed that during the 10^{-35} seconds after the big bang, an unknown force caused the universe to expand at a rate much faster than the speed of light. This expansion didn't violate Einstein's law that nothing can travel faster than the speed of light, since it was empty space that was expanding. Empty space could travel faster than the speed of light. The universe expanded at a rate of 10^{50}. After this inflation, the universe settled down and evolved according to previously accepted theory. Assuming that Guth's theory is true, the universe, which measures approximately 36 billion light-years across, originally started as a sphere a few centimeters across.

Inflation solved the three main questions about the big bang. The incredible expansion of the universe means that the regions of space that are so far apart were once very close before inflation, thus explaining their uniformity. The incredible expansion would also dilute any curvature in the universe. The region of space we can see appears to be flat, because we can't see far enough away to detect the curvature in space-time. As to the monopole problem, the rapid expansion of the universe dilutes the number of magnetic monopoles in existence. They are so rare that finding one is extremely unlikely.

The "Graceful Exit" Problem and Parallel Universes

Guth's inflation proposal solved the three great problems of cosmology, but it raised one new problem. This problem soon became known as the graceful exit problem. Once inflation started, what stopped it?

Some years after Guth's original idea, the Russian physicist Andrei Linde provided a partial answer when he proposed that spontaneous breaking was a continuous process in the universe, that big bangs took place all the time, with continuous inflation, and new universes emerged from older universes. According to Linde, we live not in a universe but in a universe of universes, a megaverse or multiverse.[16] Linde's notion suggested that our universe might have been created from a previous universe and that this process continued eternally with no beginning and no end.

Inflation combines cosmology with particle physics. Since particle physics is a branch of quantum theory, it implies that there is a "finite probability for unlikely events to occur."[17] An unlikely event would be the creation of parallel universes. In other words, quantum mechanics dictates that as soon as we assume that one universe can be created by inflation, we must assume that there is a chance that infinite numbers of parallel universes can be created.

Inflation has been shown to be consistent with all the data obtained from the WMAP (Wilkinson microwave anisotropy probe) satellite in 1998. The information gathered by the WMAP provided cosmologists with a detailed image of the universe as it existed when it was 380,000 years old. Captured on film by the WMAP was a photo of the sky displaying the microwave radiation created by the big bang itself.

Does inflation work with M-theory? Does the most startling idea in cosmology work with the most revolutionary notion in quantum mechanics? To the relief of scientists worldwide, the two theories seem to complement each other. If the universe is actually a brane, as theorized by M-theory, then it makes sense that the universe is flat. That's because branes are flat. The solutions of the other two major problems of cosmology are also solved by the nature of branes.

Both M-theory and inflation predict that our universe is not alone: that the cosmos is filled with parallel universes. According to quantum theory, there are an infinite number of these other universes, and every action and event is possible on them. Assuming that inflation is proved to be true sometime in the near future, we will finally have the answer to our question at the end of the previous section: not only are "The Mist" and *From a Buick 8* possible—they are a certainty.

LONGEVITY AND GENETIC RESEARCH

The Golden Years

Now you listen to me, you over priced, overpaid simpleton!
TIME . . . marches to a different drummer, do you not see that?

—*The Golden Years*

Once again, Stephen King supplies a new twist to an old theme, in this case, immortality. The short way of describing King's anti-aging method is to state that it requires an exploding particle accelerator. But as is always the case with King, there's a lot more to the story.

Searching for Immortality

Like many horror writers, Stephen King is concerned not only with the length of our lives but with the quality of life as we age. In the television miniseries *The Golden Years*, King looks into the horrors of growing older and the attempts by modern scientists to reverse or prevent these changes. King has a fondness for elderly characters in major roles and often uses them as the voice of reason in his most outrageous situations. Earlier in this book we discussed Ralph Roberts, the seventy-year-old hero of *Insomnia*. In the six-part serial novel *The Green Mile*, the story is told by the retired prison guard Paul Edgecomb, who is a lot older than he looks.

In *The Golden Years* (1991), the janitor Harlan Williams, who is in his seventies, rides his bicycle to work, huffing and puffing from the effort. He's not in the greatest shape, but, then again, he is a senior citizen. He arrives at his place of employment, Falco Plains Agricultural Testing Facility, where dozens of men are spraying chemicals on the lawns. The lab building where Harlan works looks like a grain silo. It's cylindrical, not particularly wide or tall, and adjoins the lawns where the men are spraying chemicals.

But inside this grain silo is a top-secret Department of Defense military operation. Research is taking place here on aging, the regeneration of lost limbs, and possibly more, all in an effort to keep soldiers on the front lines longer.

The scene shifts from Harlan to a mad scientist on the top floor of the grain silo. The scientist, Dr. Toddhunter, is alone in a sealed area, injecting something into white mice and putting the mice in an empty glass box in the middle of the room. Whirling around the top of the glass walls is a solid streak of something, which is described as a "virgin particle accelerator."

Two lab technicians are flipping switches and pressing buttons outside the secure area. One of them notices that red lights are flashing on the overrides, but the mad scientist refuses to heed the warnings. As the scene fades back to Harlan, a technician says that they must not continue because they are using an "untested process" that "requires a great deal of power."

Shift to Harlan, our lovable elderly janitor. He's being fired by Major Morland because his vision is 6 percent below the minimum required by the Department of Defense. Harlan points out that, by law, he's allowed to take the eye exam again.

At this point, we shift back to the top of the grain silo, where Dr. Toddhunter instructs the technicians to turn on the particle accelerator. The mice squeal. The room starts to glow green. And a technician cries that they must stop work now because there's "more power than a supernova" and "God help us if it goes wrong!" Smoke billows from the glass box holding the mice. The mice are green, now greener. Dr. Toddhunter is quivering and going mad. And then the room blows: the equipment sparks, flames, and explodes. Glass shatters, and the particle accelerator blows up.

In the hallway right outside the lab, Harlan hears the smoke alarm go off, and he grabs a small fire extinguisher. But as he runs toward the lab, Dr. Toddhunter races past and slams into him, and Harlan falls. The two technicians attempt to quell the explosion using a fire extinguisher, and as Harlan revives and tries once again to race toward the lab, a huge explosion knocks him down in a wash of green glow.

Harlan survives, but the technicians die. Toddhunter seems fine, although he remains arrogant and crazy. The military men explain to the investigating blonde female ex-CIA security chief (who wears tight skirts and high heels) that K93 is the by-product of a chemical reaction and is the green glowing stuff seen everywhere during the explosion. It is what will help to heal people and regenerate tissue.

Meanwhile, the meat of the story and the subject of this chapter: Harlan starts getting younger. His white hair turns brown. His eyesight is superior. He's physically about fifty, no longer in his seventies.

We can speculate about all sorts of things related to Harlan and his new youth. How will he react to his wife when she's seventy-five and he's only twenty-five, then twenty, then ten, then two? Is it possible for someone's age to drop until he reaches the zero mark and moves into the fetal stage? How will he get shorter, how will his brain attain the size and the composition of an infant? As he gets younger and younger, will he forget things that he learned in his older, now-gone years? Will the wisdom of his old age be lost forever? King hints at the answers to these questions.

Exploding Particle Accelerators and Aging

Now, it seems unlikely that a particle accelerator would be used in a grain silo with an open glass box of, say, a dozen mice. It seems unlikely that injecting the mice with chemicals of some kind and then turning on a particle accelerator would reverse the aging process in the mice. The amount of power generated by the accelerator would kill the mice, for one thing; and for another, the mice are in an open glass container, with a lone scientist—who is not shielded in any way whatsoever, not even with lab goggles—mere inches away from them. Why wouldn't the scientist be killed by the dose of power from the accelerator as it supposedly knocked into the mice with shocking levels of power ("more power than a supernova")? On a related note, how can the accelerator possess more power than a supernova? When the accelerator blows up, why isn't everyone killed? How do Dr. Toddhunter, the security chief, the military guys, and Harlan survive at all? And even the mice survive, and they were in the room getting blasted by the accelerator! At any rate, there are clearly some logical flaws in the science as presented, but the series makes for excellent television and is highly entertaining and thought provoking.

Let's set aside the issues of the exploding accelerator, the green goo, and so forth, and focus for a moment on the process of aging. If we know something about the aging process, it will help us understand how—if at all possible—the aging process could be reversed.

Aging is defined loosely as the deterioration of nearly all body functions over time. In the United States, the average lifespan is now more than seventy-five years, and this is largely due to better medical and health resources, sanitation during the preparation of food, and better water quality.

There are pockets of the world where lifespans greatly exceed those of the so-called advanced countries. For example, in the Karakoram Mountains, the Northern Andes, and very isolated areas of the Caucasus, people tend to live for a hundred years or more, retaining excellent health. These areas do not have good medical and health resources, sanitation during the preparation of food, and safe water quality. In fact, by most standards the people there lead pretty harsh lives. They are hit by infectious diseases for which they have no cures. They have high infant mortality rates, they are illiterate, and they lack modern sanitation facilities.[1] So why do they live so long? That's something that scientists have been trying to discover for the last quarter-century. And, with any luck, they'll know the answer sometime soon.

Let's first take a look at immortality in science-fiction literature. From there, we'll move to genetics and see how close we are to making that fiction fact.

The Two Faces of Eternal Life

Stories about the quest for immortality are as old as civilization. The oldest surviving legend, the Sumerian Epic of Gilgamesh, from approximately 2700 B.C., tells of an all-powerful king seeking the secret of eternal life in a far-off country, finding it, and then losing it.

In the *Odyssey* by Homer, Odysseus is offered immortality by his lover, the demigoddess Calypso, but he rejects it because he wants to return home to his human wife, Penelope.

In *Gulliver's Travels* (1726) by Jonathan Swift, Gulliver meets, during the course of his voyages, the Struldbruggs, humans who live forever. Thinking that immortality would be a blessing, Gulliver is dismayed to learn that the Strudlbruggs are the most obnoxious and ugly beings imaginable. The older the Strudlbruggs get, the less they can remember. Nor are they spared the ravages of time. Their bodies continue to age even though they remain alive for hundreds and hundreds of years. Their only passions are envy for the vices of youth and desire for death.

By the end of the nineteenth century and the beginning of the twentieth century, novels dealing with immortality or eternal life had separated into two distinct branches. In the pessimistic vein were stories that treated living forever as a curse that would bring only unhappiness and despair. Typical of these sorts of novels was *After Many a Summer Dies the Swan* by Aldous Huxley (1939).

The other viewpoint on immortality was that eternal life would remain eternally interesting. Championing that view was the Wandering Jew series by George Viereck and Paul Eldridge, which began with *My First Two Thousand Years* (1928). In the novel, the hero, who is cursed by Christ to wander the earth until the Second

Coming, meets just about every famous historical personage in history during the course of his travels around the world, living life to its fullest.

In the science-fiction magazines of the 1930s, both sides of the immortality debate were present. In the Professor Jameson series by Neil R. Jones, beginning with "The Jameson Satellite" (*Amazing Stories*, 1931), the brain of a scientist is discovered intact in space 40 million years in the future by a race of creatures known as Zoromes. The Zoromes have achieved immortality by transferring their brains into near-indestructible metal bodies. They do the same for Professor Jameson, who then joins them on their unending quest for knowledge throughout the cosmos.

In Laurence Manning's five-part series *The Man Who Awoke* (*Wonder Stories*, 1933), the banker Norman Winters, disillusioned by the events of World War I, goes into suspended animation for five thousand years at a time in hopes of discovering the meaning of life in the far future. Winters never succeeds in finding a purpose, but he does achieve immortality in the last episode of the series, "The Elixir" (*Wonder Stories*, August 1933), so he can keep on looking.

In Dr. David H. Keller's novel *Life Everlasting* (*Amazing Stories*, January–March 1934), a scientist discovers a serum that will bestow eternal life on the user. The drink also wipes out all criminal impulses and makes people happy. But it completely destroys all sexual impulses and the desire to have children. The race is immortal but unchanging. Faced with the choice of immortality or having children, mankind selects the latter. The serum is discarded and life returns to normal.

Science fiction never clearly resolved the argument between the two sides of the immortality debate. For every story that appeared arguing that immortality would be a blessing, another was written saying that it would be a curse. In the 1940s and 1950s, a number of novels and short stories take no side in the argument but instead feature immortal heroes so busy saving the world or some part of it that they have no time to get bored.

Typical of the time was A. E. van Vogt's *The Weapon Makers*

(1951), in which the immortal hero, Robert Hedrock, must deal with a thousand-year-old monarchy and a guild of weapon makers who oppose it, both of which he created in the distant past. While managing that, he must also pilot the first interstellar spaceship to a landing on an Earth-type planet many light-years from Earth— and negotiate with a race of interstellar bugs.

Hedrock is portrayed as an altruistic immortal, a man who by unusual circumstances has become immortal and feels an obligation to keep the human race alive and well. A similar character, Conrad, is the lead of Roger Zelazny's *This Immortal* (1966). Immortal heroes are featured in Robert A. Heinlein's *Methuselah's Children* (1959), Wilson Tucker's *The Time Masters* (1953), and Clifford Simak's *Waystation* (1963). Immortal villains play prominent roles in *Minions of the Moon* by William Grey Beyer (1950), *Carrion Comfort* by Dan Simmons (1988), and *Phantoms* by Dean Koontz (1981).

During the last decade, science fiction has treated extended life and immortality as one of the direct results of modern-day technology. Novels by such writers as Peter Hamilton and Alastair Reynolds feature numerous characters who are hundreds of years old. Immortality is assumed to be one of the basic facts of life in novels of the future, leading us to wonder, how close is eternal life? Is science going to keep us eternally young?

The Science of Immortality

It's possible that genetics is an important factor in longevity. We talked about genetics quite a bit in chapter 1, specifically when discussing Charlie McGee in *Firestarter*. We wondered whether her parents' genes really could have been mutated, hence giving her the ability to set fires simply by using her mind. Let's look more closely at genetics, with the objective of determining whether aging is caused by genes, and if so, whether aging can be reversed by genes.

A Genetic Cure for Diseases

Experts predict that by 2020, all human diseases and disorders will be linked to the human genome. If we test for the diseases and the disorders, then over time we'll stand a good chance of eliminating most of them. Indeed, the futurist Michael Zey reports:

> The genetic approach to disarming and neutralizing diseases is moving ahead at a breakneck speed. The Human Genome Project is the first stage in this battle—identifying genes and their functions. The second stage, which some think will achieve total success by 2020 or earlier, involves delivering these genes into the human being to correct both the maladies themselves and an individual's vulnerability to developing a specific disease.[2]

According to the *New York Times*, we already know the genetic foundations of 90 percent of all breast cancers. If we know what triggers breast cancer, we can either prevent the disease or help to cure it by fixing the genes.[3]

If we eliminate a lot of fatal diseases, it follows that we'll live longer. Some experts predict that the average lifetime of a person will increase to 105 by the year 2025. More conservative estimates place us at 85—even so, a dramatic increase over our average age at death today.

If we crack the genetic instructions that enable us to reverse or stop aging, we'll live even longer, possibly becoming immortal, or so one might assume.

Telomerase, the "Immortality Enzyme"

In 1997, *Time* magazine wrote that doctors at the Geron Corporation and the University of Colorado at Boulder discovered what they called an "immortality gene."[4] Of perhaps more interest was the discovery of the telomerase enzyme and the role of telomeres in aging. In the mid-1990s, researchers learned that the telomeres, which are the tips of chromosomes, shorten every time cells divide and therefore replicate.[5] After approximately fifty replications, cells stop dividing, beginning what scientists believe is the aging process. Scientists have long felt that if they could lengthen the telomeres, then the cells would continue to divide, and hence people wouldn't age as fast. The flip side, of course, is that if the cells divide too much, then people get cancer.

In 1998, scientists discovered telomerase. It was predicted that the enzyme could immortalize cells.[6] Significantly, in January 2000, scientists at the Geron Corporation in Menlo Park, California, succeeded in lengthening the telomeres in cells by activating the telomerase enzyme. They were able to lengthen the lifespan of cells, triggering another twenty cell divisions past the normal fifty.[7]

Zey claimed that if "we could find a way to keep our telomeres intact, we could theoretically live forever. When [telomerase] is expressed in a cell, that cell is for all practical purposes 'immortal'; many biologists have dubbed telomerase the 'immortalizing enzyme' because of its power to bestow 'life everlasting' on cells."[8] Furthermore, Zey explained that the telomerase technique "will not enable a 40-year-old man to regain his 20-year-old body; however, it will prevent him from aging past 40, because his cells will not age or die."[9]

In simple terms, a chromosome is a double-stranded, twisted molecule of DNA that resides in the cell nucleus. Our genes are contained in the chromosome. Being part of the chromosome, the telomeres are also made up of DNA sequences of the four nucleic acid bases: G (guanine), A (adenine), T (thymine), and C (cytosine). In vertebrates, telomeres consist of the repetitive DNA sequence TTAGGG on one DNA strand that is bound to the repeating sequence AATCCC on the other DNA strand.

While a chromosome has approximately 150 million base pairs, the human blood cell's end structure—the telomere—has about 8,000 base pairs at birth. As we age, our telomeres become much smaller, containing perhaps 3,000 to 4,000 base pairs. In elderly people, telomeres have only 1,500 base pairs.

As noted, the typical cell divides about fifty times, and each time a division occurs, the cell's telomeres shorten. In fact, each shortening deletes from 30 to 250 base pairs from the telomere. Telomeres that are approximately 10 to 12 kilobases long enable cells to divide about fifty times. In computer science, a kilobit is 1,000 bits (1,024 bits to be precise). In genetics, a kilobase is 1,000 base pairs of

DNA. When telomeres are 4 to 6 kilobases long, cells no longer divide, and they die.

If there were no telomeres, then the part of the chromosome containing the genes that are the basis of our existence would shorten instead. Without telomeres, cell division would delete essential genes from our bodies. Without cell division, our bodies could not create new blood, bone, skin, and other cells. Our cells would become cancerous or die.

In addition, if not for telomeres, cells might interpret the ends of chromosomes as broken DNA. In this case, the cells might attempt to fix the broken DNA, causing the cells to malfunction and die. Cells "fix" broken DNA in two ways: homologous recombination and nonhomologous end-joining. The first method, homologous recombination, is free of errors, but it needs the homologue chromosome of the pair to work. Nonhomologous end-joining, while it does save the chromosome from further damage, often produces errors in the genetic material. The telomeres prevent the cells from using nonhomologous end-joining to fix chromosomes.

For a cell to divide, its chromosomes must first be duplicated so that the new cell will contain identical genetic material. The duplication process begins when the two strands of the chromosome separate from each other, and then an enzyme called DNA polymerase works with short pieces of primer RNA to make two new identical strands. The primer attaches to the DNA strand, enabling the cell to copy the DNA material. After all of the DNA material is copied and the new strand exists, the primer does not attach to the end of the new strand. Hence, the new strand is shorter—specifically, by 30 to 250 base pairs—than the original strand. When the cell divides again, the new copy of DNA loses more of the end section. When it divides a third time, even more end material is lost, and so forth. The telomeres protect the actual genetic information on the DNA strand from being lost during cell divisions.

The telomerase enzyme, which exists in egg and sperm (germline) cells, adds bases to the ends of the telomeres, making them "immortal," to some extent. The telomerase does not become depleted in the germline cells, which pass from one generation to the next. Without telomerase to protect the length of germline telomeres, humans would become extinct.

In other (somatic) body cells, the telomerase may not be expressed at all. In particular stem cells, which we discuss later in this chapter, telomerase is expressed but regulated tightly. In 80 to 90 percent of tumor cells, telomerase is expressed. A cell that is beginning to become cancerous divides more frequently than a normal cell does. To protect itself from dying due to short telomeres, the cell becomes a cancer cell and activates telomerase to keep its telomeres from becoming too short. This type of cell can keep dividing. As an aside, we may be able to diagnose cancer by measuring telomerase in cells, and, by extension, we may be able to cure cancer by detecting a way to stop the production of telomerase in cancer cells.

Oxidative Stress Contributes to Aging

Telomere shortening is not the only cause of death. Keeping telomeres long may increase our lifespans, possibly by twenty or thirty years, but there are other reasons why people age and die.

For example, oxidative stress, which is caused by the metabolism of oxygen, damages our DNA, lipids, and proteins. Highly reactive oxidants, which contain oxygen, are created when we breathe. They are also created when we smoke cigarettes, drink alcohol, become infected, or develop a tissue inflammation. An oxidation reaction transfers electrons from one substance to an oxidizing agent. The amount of oxidative stress is based on the rate at which the damage is induced in the body against the rate at which the damage is repaired. The damage is induced when oxidants are generated in the body, and the damage is repaired or eliminated when antioxidants inactivate the oxidants. According to biochemists and molecular biologists at the University of California, "Oxidant by-products of normal metabolism cause extensive damage to DNA, protein, and lipid. . . . This damage (the same as that produced by radiation) is a major contributor to aging and to degenerative diseases of aging such as cancer, cardiovascular disease, immune-system decline, brain dysfunction, and cataracts."[10]

Antioxidants exist normally in all living cells. They reduce the rate of oxidation reactions and prevent chemical damage to the cells. While a lot of research and contemplation has occurred around the use of antioxidants to keep us healthier as we age, the debate is still open as to whether antioxidant supplements and reduced calorie intake truly add years to our lives. For now and in the imminent future, oxidative stress remains a contributor to the aging process.

Sugar Blues

Another factor in aging is glycation, which occurs when a sugar molecule such as glucose or fructose binds to DNA, lipids, and proteins, rendering them unable to function correctly. Though we may not be happy without our doughnuts and muffins, the less we eat, the better off we'll be.

Replaceable Body Parts

In addition to providing gene therapies that keep us from aging and that regenerate our body organs, doctors are working on ways to regenerate damaged limbs and skin. This research will also extend our lives, and if all of these methods work, it's conceivable that by 2050 or so, we may very well live to 150 or more. If you think we're being overly optimistic, don't forget that various experts—scientists and medical doctors—predict that we may actually become immortal.

We've long used heart pacemakers and artificial knees, hips, and other contrived joints to keep our bodies going. A lot of people wear glasses and contact lenses. We have hearing aids and false teeth. It's accepted practice that we put foreign matter into our bodies to enhance our health and extend our lives.

It's not a far stretch to imagine that soon we'll have artificial hearts made from batteries, pumps, and pistons. In fact, five U.S. hospitals are experimenting with artificial hearts that are operated by microprocessors and internal sensors, and doctors predict that sixy thousand patients a year will have artificial heart implants.[11]

As for tissue regeneration, you may have seen the video clips of the University of Massachusetts and MIT experiments that produced the mouse with the human ear on its back. The human ear mouse and such creatures as custom-designed pigs are used as "animal crops" to supply body organs for human transplant.

Tissue regeneration is common in many invertebrates, which detach their own body parts and sprout new ones. When injured, an invertebrate may regenerate tissue and limbs. Examples of invertebrates that regenerate tissue and limbs include planarians, or flatworms, that can be cut into fifty pieces, resulting in fifty distinct individual worms. Also included are crickets, which regenerate their legs, and starfish, which regenerate lost arms. Invertebrates as a whole make up 97 percent of all animal species, with the vertebrates being mammals, birds, fish, reptiles, and amphibians. Common invertebrates are earthworms, roundworms, flatworms, sponges, jellyfish, starfish, sea urchins, sea cucumbers, insects, spiders, grasshoppers, crabs, lobsters, snails, and squid.

Some vertebrates can regenerate lost limbs and tissue. Lizards, for example, can detach their tails from their bodies to distract predators. The predators eat the tails, the lizards escape, and later, the lizards take their time regenerating their lost tails. The new tails are never quite the same as the original tails, and if a lizard loses a leg, it cannot regenerate the lost limb. Tadpoles can regenerate lost tails, but frogs cannot regenerate tails or limbs.

Even humans can regenerate tissue to some degree. We generate skin and finger nails, as well as blood vessels and peripheral nerves under particular conditions.

Most vertebrate embryos—and some adult animals—are able to regenerate limbs and damaged nerve connections. The newt and the axolotl, both salamanders, and the hydra can regenerate limbs and nerve connections as adults. The urodele amphibians, which include the newt and the axolotl, are the only vertebrates that can regenerate extremely complex structures, such as eyes, jaws, spinal cords, tails, and limbs.

As soon as a salamander loses a limb, it begins to regenerate it. Within twelve hours after an axolotl loses a limb, skin migrates over the area where the limb has been lost. An epidermal cap, or stump, is formed. Within days, the stump tissues form a blastema, a mass of dedifferentiated proliferating cells, and genes are activated that were initially used when the limb was first created in the embryo. Dedifferentiation means that cells lose their specialization and return to a more basic form. The first part of the new limb to be created is the hand or the foot, also known as the distil tip of the limb. Eventually, the regenerated limb contains blood vessels, muscles, and motor neurons, and connections are established that make the limb function normally. It takes from one to three months for the new limb to form completely and to operate normally.

As for mammals, tissue regeneration includes fingertips, as mentioned earlier, as well as antlers and holes in ears. If the membrane around a human rib remains intact, the rib itself can regenerate. Adult mammalian skeletal muscle also regenerates. The MRL mouse, another mammal, has extraordinary regeneration capabilities and can even heal surgically induced cardiac wounds within sixty days.[12]

Doctors have discovered ways to grow and use artificial human skin. For example, a company called Advanced Tissue Sciences has a product called Dermagraft that repairs burned skin.[13]

The *New York Times* reported:

Regenerating a limb could be, in the next decade or two, a mere assembly job of coordinated parts—muscles, bone, skin—grown in vitro, seeded upon scaffolds and stimulated by growth factors. But such methods will seem crude in comparison with what the genomic revolution will eventually make possible. Because the programming instructions for growing a human body reside in every human cell, in the form of 46 chromosomes that contain roughly 100,000 genes, most cells, properly stimulated, could set the regrowth of a limb in motion. Scientists will first have to parse the sequence of gene activation and protein expression that triggers tissue growth from progenitor cells. The next step will be identifying the switches that coordinate limb development. Knowing when, and in what order, to pull the levers will allow humans to recapitulate development.[14]

If we can grow human ears on mice, then we can grow human fingers on human hands.

In fact, the biotechnologies of tissue engineering will be major factors in our increased longetivity. "A basic goal of tissue engineering," wrote Robert M. Nerem, an institute professor and Parker H. Petit Distinguished Chair for Engineering in Medicine at the Georgia Institute of Technology, "is to fabricate living tissue equivalents, that is, biological substitutes to be implanted into the body." He also noted that the development of materials is critical to "promote the remodeling of tissue." Furthermore, he discussed various applications of tissue engineering, among them: artificial skin, blood substitutes, bioartificial organs, neurological implants, orthopedic devices, and vascular grafts.[15]

The sum result is that if we reduce the aging of our cells by using telomerase therapies, replace our worn body parts and organs, and prevent most deadly diseases, we will live much longer than we do today.

Unforeseen Quandaries

Whether anyone wants to be age 150 while retaining a 20-year-old body is questionable. At 70, Harlan from *The Golden Years* is a lot different from the young Harlan of Coney Island. His mind has matured in many ways. His older body is in tune with his older mind. His older body can't be replaced with a younger version and still make sense to his older mind.

The social problems involved with extending our lives will definitely affect dating, marriage, love, and sex. Younger people will have a much harder time finding one another in an elderly world. For example, it will be even more difficult to find a decent job if people are working until they're eighty years old; it will mean that products and advertising will shift from the ultra-young dramatically toward older populations, who will have the lion's share of buying power. What will it be like to be an eighty-five-year-old engineer reporting to a twenty-five-year-old employer? Will age discrimination get worse, or will it reverse? If Harlan remains a janitor, yet is a seventy-five-year-old man trapped inside a sixteen-year-old body, who will want to hire him? How will his bosses treat him?

If we can limit the aging of our bodies, then we must figure out a way to limit the aging of our brains. After all, the brain is part of the body. By keeping your body at age twenty, would you also retain the attributes of your twenty-year-old brain forever? If so, you would forfeit the wisdom, the skills, the knowledge, and the emotional transitions that come with age.

In addition to personal questions about extending lifespans to 150 years or more, we must also ponder environmental, legal, and other basic issues. For example, if everyone is living to 150, our elderly population will rise to unprecedented levels. We'll have very few young people working to support the huge new population of the elderly. Who will pay for the health insurance, and who will pay the medical bills of a population that's perhaps 60 percent (or more) retired? Decades from now, the percentage could shift so dramatically that 80 percent of the population is retired, with only 20 percent still working. The older population will require genetic therapies to extend their lives: who will pay for all these medical services? Will we even have enough doctors and trained paraprofessionals to handle this huge explosion in requested genetic therapies?

If our population does indeed rise due to longer lifespans, will pollution also increase, as well as housing and food shortages? Will we have sufficient resources to handle a massive explosion in population on planet Earth?

Nanotechnology to the Rescue

Eric Drexler, famous for his work with nanotechnology, advanced some extremely positive arguments in favor of increased lifespans in his book *Engines of Creation*. In a section titled "Long Life and Population Pressure," Drexler argued that nanotechnology will give us cell-repair machines that not only extend our lives but also help us heal the earth. Any damage we do to the planet, he concludes, will be fixed by the very nanomachines that fix our bodies. He wrote:

[As] cell repair machines extend life, they will increase population. If all else were equal, more people would mean greater crowding, pollution, and scarcity—but all else will not be equal: the very advances in automated engineering and nanotechnology that will bring cell repair machines will also help us heal the Earth, protect it, and live more lightly upon it. We will be able to produce our necessities and luxuries without polluting our air, land, or water. We will be able to get resources and make things without scarring the landscape with mines or cluttering it with factories. With efficient assemblers making durable products, we will produce things of greater value with less waste. More people will be able to live on Earth, yet do less harm to it—or to one another, if we somehow manage to use our new abilities to good ends.[16]

In brief, nanotechnology operates at the nanometer scale of molecules and atoms. This type of technology will function with the precision and the basic invisibility of molecules. It was Richard Feynman in 1959 who first expressed this vision in a talk called "There's Plenty of Room at the Bottom." Feynman and later Drexler had the early vision that machines would someday exist at a nanoscale level, and that these machines would be able to build atomically precise, digitally controlled equipment.

When Drexler came up with the term *nanotechnology*, he was referring to a world much smaller than the one based on microtechnology, which is any technology that operates at the micron level, where a micron is one millionth of a meter. Most commonly, we think of microtechnology as the photolithographic chip-making process for computers. Nanotechnology would manipulate matter at a much smaller level than microtechnology does; it would manipulate matter at the nanometer level, where a nanometer is one billionth of a meter. Hence, 0.05 microns is the same as 50 nanometers. A reasonable definition of nanotechnology might be that it deals with technologies that operate at or below 100 nanometers, and that it deals with technologies that create machines that have precise atoms and chemical bonds.

In today's world, we do see a lot of research in nanoscale technology, which means we are able to use scanning probe microscopes to view things at the atomic scale; and we can produce molecules with atomic precision, but only by mixing chemicals and doing other things that are beyond the nanoworld landscape. While molecules can be created with precise atomic arrangements, they are randomly placed with no intermolecular cohesion. In other words, we're not making nanotech devices yet.

Eventually, we will be able to take our atomically precise molecules and fuse them using self-assembly. The molecules will combine in ways that make sense and will create larger, logical things. The molecules will be built in such a way that when jumbled together, they will self-assemble into these larger objects that have a predetermined structure. Some smaller molecules and specially designed DNA molecules are already able to do this self-assembly, but much work remains in this area.

After self-assembly is achieved, the next step in nanotechnology will be to combine the molecules into machines that can create other machines. This will basically make the nanotech devices self-replicating, which means they will be lifelike. By extension, we note that should these self-replicating nanomachines obtain artificial intelligence, they could be considered artificial life.

When the self-replicating nanomachines are fairly advanced, they will become relatively common and inexpensive. In the far future, nanotechnology will exist that creates new parts from atoms rather than from nanomachines.

Circa 2007, most nanotechnology extends current manufacturing and materials science. In the case of nanotechnology, the processes simply involve things that are shorter, thinner, or smaller than 100 nanometers. Ultrafine nanotech powders, for example, can be purchased in grocery stores today.

Sugar consists of groups of atoms, or molecules, that are held together by covalent bonds. The molecules are held together by forces that are much weaker than covalent bonds. When sugar dissolves in hot liquid, such as tea or coffee, the bonds between the molecules break, and the molecules each float separately in the liquid. If you then pour milk into the hot liquid, the blobs of milk fat don't dissolve, yet they are small enough to become suspended, or dispersed, throughout the liquid. This is commonly called a colloidal suspension, and particles that are a billion times the volume of a sugar molecule can form this type of suspension. The finer the particles, the more evenly dispersed they will be in the liquid. If the particles are of a nanosize, the suspension will be very even, and the properties of the resulting mixture will be more stable and easier to predict and manipulate. Some ultrafine powders available today are sold as high-tech cleaning solutions and stain removers.

During the 1990s, the buckytube, a carbon nanotube, made a big splash in the press. In short, a buckytube is a graphite strip rolled into a seamless cylinder. Graphite happens to be extremely strong, and in the shape of a seamless cylinder, it is very difficult to rip. Another useful property of the nanotube is that it is slippery, and if you put one tube inside another, you can slide and rotate them easily. To build things out of the tubes, engineers often use special additives to decrease the slippery nature. Some buckytubes are employed as semiconductors, others as conductors, and buckytubes have been used to make both diodes and transistors.

Today's microprocessors require transistors and wires that are less than 100 nanometers in size. So the lithography that creates microchips is a nano process. While we can make micromachines, we have yet to make nanomachines using lithography. Progress is being made in this area, however, with companies already working with microlithography to build micromanipulators that assemble other micromachines.

One realm of advanced nanotech research is the ability to read and write data at the molecular level. It's been twenty years since we learned how to read data at the atomic level, but the writing of the data has proved more tricky to accomplish.

Research is most advanced in the work being done with nano-electronics, with molecular switches, wires, and memory being developed. Engineers are working to chemically create switches and wires and then to attach these resulting components to DNA. If the DNA self-assembles, it will make the molecular electronics into working circuits.

Drexler and other scientists envision a time when nanomachines will destroy cancer cells and viruses. These nanodevices will clean our bodies, monitor and repair themselves, and self-replicate. They will be tomorrow's age-extending drugs.

By the time we've significantly increased our ability to prolong life with genetic therapies, nanotechnology will still be in its infancy, and cell-repair machines that cure the earth will not yet exist.

Stem Cells: A Future Panacea?

Along with telomeres, tissue regeneration, and nanotechnology, another area that promises to extend our lives is research related to stem cells. Currently, stem cell research is under attack by many people, while others think it holds the key to our future.

Stem cells are able to develop into many kinds of cells. The idea is that stem cells can divide endlessly and replenish the body with all sorts of other cells. Each new cell can remain a stem cell or become a specialized cell—a muscle cell, a brain cell, a red blood cell. If tissue regeneration is needed, perhaps stem cells can do the job. If someone is suffering from Lou Gehrig's, Parkinson's, or other neuronal diseases, maybe stem cells can help by replacing dead and damaged motor neurons with new normal ones. Stem cells may be able to generate replacement cells for injured heart cells, hence giving us a cure for heart disease. One day stem cell therapy may even cure diabetes.

There are two types of stem cells being used for research: embryonic stem cells and adult stem cells. The intense public debate about stem cell research focuses on the former—in particular, on the use of human embryonic stem cells. It's beyond the scope of this book to delve into the ethical and moral debates surrounding this issue or to comment on the creation of embryos through in vitro fertilization, the use of aborted embryos, and so on. Our focus is on science related to Stephen King's novels and, in the case of this chapter, the science related to longevity.

More than twenty years ago, scientists were able to obtain stem cells from mouse embryos. Then in 1998, after years of intense research, scientists discovered how to isolate stem cells from human embryos and how to cultivate and grow cells in laboratories.

In a blastocyst, or an embryo that is only three to five days old, stem cells start to form specialized cells that will eventually make up human organs, skin, and other tissues. In adult humans, bone marrow, the brain, and the muscles contain stem cells that generate replacement cells for those that are damaged or diseased. Stem cells that differentiate into specific types of other cells can become a major source of replacement cells and tissues.

It is the ultimate hope of scientists that stem cell research will lead to therapies for spinal cord injuries, burns, strokes, rheumatoid arthritis, Parkinson's disease, Alzheimer's disease, heart disease, diabetes, and Lou Gehrig's disease, among many others. Researchers hope that we might be able to revive any organ in the body with a simple injection of embryonic stem cells.

As noted, the reason this might be possible is that stem cells can turn into other types of cells. But how can these new cells divide more than fifty times before they themselves die? After all, their telomeres will shorten long before they can create enough tissue to replace, say, a lost limb. To divide enough times to cure a neuronal disease, cells need telomeres that last. With revived telomeres, these stem cells—someday, we hope—will multiply and resuscitate damaged body organs, giving us cures for diseases and much longer lives.

Limiting Doughnuts and Muffins

Harlan's story is a fun and fascinating one, but we hope that Dr. Toddhunter's green glowing particle accelerator never becomes a reality. We'd rather see humanity turn to telomere research, tissue regeneration, nanotech, and stem cell research, and to realistic methods to control oxidative stress and glycation, rather than expanding our lifetimes with blasts from particle accelerators. It might make more sense—at least, for now—to limit the doughnuts and muffins.

EVIL, OBSESSION, AND FEAR

The Tommyknockers • *Carrie* • *The Talisman*
It • *The Stand* • *Danse Macabre*
The Shining • *Misery* • "Night Surf"

Redrum. Redrum. Redrum .

—*The Shining*

Stephen King is a master at making us feel the presence of evil. We rarely see this evil in the grotesque forms that other horror writers use to portray it. In King's work, we hardly ever see ghouls, vampires, or dripping, bloody, disemboweled killing monsters. We don't see cannibals eating flesh or gigantic man-eating octopi and sharks. There are no shower scenes in King's movies; no young girl is naked and lathered in soap, only to turn toward the camera, displaying her breasts and shrieking in horror as a knife swoops down or a raging killer descends upon her. King is more subtle, which is why his stories are so powerful and unforgettable.

See No Evil

King and Alfred Hitchcock share the role of scaring us psychologically rather than through monsters and gore. Like King, Hitchcock used the common man and woman to strike fear in us—innocent people who find themselves in circumstances way beyond their control and understanding.

For example, in Hitchcock's *The Birds*, flocks of raging birds descend upon a small town and attack the people. Yet Hitchcock never tells us why the birds are attacking humans. We assume that something has aroused them to act, something that the innocent Melanie has done unintentionally. They attack only when she arrives in Bodega Bay, which we learn through a hysterical mother's dialogue in a restaurant. Melanie comes to town to pursue a love interest, Mitch, bringing him caged lovebirds as a gift. It's possible that Hitchcock used the raging birds to symbolize various aspects of human behavior, such as sexual flirtation and tension, unrequited love, or a small town's inability to be friendly to strangers. When the humans finally escape from Bodega Bay, they take the benign lovebirds with them. It's possible that the birds are simply evil—whatever that means—but unlikely.

In Hitchcock's *Vertigo*, the detective Scottie has a terrible experience with heights that put him in a state of vertigo. A friend asks Scottie to spy on his suicidal wife, Madeline, and Scottie finds himself falling in love with her. Scottie becomes obsessed with Madeline, and when he cannot overcome his fear of heights to save her life, he has a nervous breakdown. Eventually, Scottie recovers and meets Judy, who resembles the dead Madeline. Still obsessed with Madeline, he makes Judy change her appearance so that she looks like his dead lover. *Vertigo* is about obsession and fear.

Evil, obsession, and fear: Hitchcock was a master of all three—and so is Stephen King.

In King's novels, innocent people come face-to-face with evil all the time. Sometimes it's in the form of kids and adults taunting outcasts. Sometimes it's more powerful, in the form of bullying. And sometimes it's in the form of serial killers and tyrants.

As with Hitchcock, King's innocent people often lose to evil forces, not because the people are bad or because they are villains, but precisely because they are naive. Because King's evil isn't as brute force and visible as a ghoul, a vampire, or a man-eating shark, his characters may not even recognize evil when they see it. Or they may encounter evil, as in *The Tommyknockers*, without knowing how to resist and defeat it. They might even try really hard to defeat evil when it confronts them but find that they are too weak to overcome it. In the end, evil usually wins in Stephen King's novels.

In *Carrie*, evil is in the form of thoughtless teenagers who bully a naive girl. A lack of empathy and kindness reaps consequences that nobody can predict or control. Pranks against Carrie lead to active hostility, and given that Carrie has tremendous powers that can destroy anything simply by thought, the teen kills everyone around her. In the end, fear takes the place of evil, and in the book (but not the movie), much of the narrative is told in the past tense by people who were there: people who remain frightened and obsessed by what happened.

After *Carrie*, King often remained true to his original theme: that bullies who torment children create evil that has horrible consequences. For example, in *The Talisman*, which King wrote with Peter Straub, adults fail to see and fight evil, while children lead the battle and try to save everyone. Here, a twelve-year-old boy enters an alternate universe to rescue his mother, who is held captive. If he doesn't go on this adventure to save his mother in the alternate world, he will not succeed in saving her in reality, and both worlds will be destroyed.

In King's novel and movie *It*, adults cannot understand the evil around them that threatens to destroy their world. While the adults see signs of evil, they ignore the signs. The children, on the other hand, not only see the evil, they believe it exists and that something must be done to save everyone from it. The monster It beneath the town of Derry emerges from its underground lair every now and then to feed, and it is almost defeated—almost—by a group of small children, all of whom are outcasts. The children feel weak, incapable of destroying It. But to them, the monster represents just one more bully. They wound the monster but must return years later as adults to kill It. During the interim years, from the time they are children fighting bullies to adulthood, they remain afraid of and obsessed with It. Once again, we find evil, fear, and obsession as the main themes in King's work, just as we observed them in Hitchcock's work.

In King's *The Stand*, the world is crippled by disease, and King's survivors are drawn toward two communities. One is led by an old woman in Nebraska, who seems to represent God, and her followers include a down-to-earth Texan, a philosophy professor, a young pregnant woman, and the pregnant woman's rejected friend. The other is the Dark Man, Randall Flagg, who represents evil, and his followers are all thugs and maniacs, including the demented Trashcan Man. The people in *The Stand* want to follow the path of least resistance. They are afraid of the Dark Man, but they are neither good (Nebraska/God) nor evil (Dark Man/evil). If good is rewarded, the people will behave nicely. If evil is rewarded, the people will behave poorly. The Nebraska/God camp sets up a democratic government. Meanwhile, the Dark Man/evil camp beats and kills anyone who doesn't cooperate with what they demand. The innocent survivors become obsessed with finding the Dark Man and destroying him, and many people are sacrificed in the long battle.

A World without Walls

Humans have always lived with conflicts, aggression, and territorial disputes. From the earliest times, some six thousand years ago, we have banded in units to survive the forces of nature because, after all, the power of several outweighed the power of one. Later, the power of many replaced the power of several. Mankind's attempts to coexist with nature have persisted and always will remain. If anything, we are in a crisis now whereby we are hurting nature rather than vice versa, and if we don't stop assaulting the environment and the world around us, Earth itself will no longer be able to sustain us.

As humans banded into larger tribes, then into states and countries, our struggles with nature were joined by our struggles against one another. Wars over territory, food, mates—all the attributes of survival in the animal world—took hold, and mankind fought itself in massacre after massacre.

Rules, regulations, and laws are created in an attempt to govern our actions. Every alliance has its military and police forces. These rules, regulations, laws, and military/police forces constitute a world with walls. We cannot seem to live without these walls.

Why can't we live in a world without walls? Are humans incapable of living peacefully, in harmony, without ill will toward one another? Are we, on some fundamental level, evil? Or are we basically good, but circumstances make us behave in what could be called evil ways?

The Concept of Evil

In most modern horror fiction, evil is portrayed as something that exists as an entity separate from people. This evil, be it in the form of vampires or giant man-eating tarantulas from the planet Zoloon, is an actual thing. Sometimes it is a force so powerful, such as the demons in *The Omen*, *Rosemary's Baby*, and *The Exorcist*, that it literally consumes people and takes over their minds, hearts, and souls. This supernatural force of evil is not scientifically based. It is a myth.

If demonic pure evil exists outside of us, literally, driving us to do terrible things to one another, then this implies that we are not responsible for our poor behavior and bad actions. "The devil made me do it," in other words.

Theological arguments that try to reconcile the existence of evil in our world with the assumption of a peaceful, benevolent God are called a *theodicy*. The word comes from the Greek words "the justice of God" and was first used in a 1710 essay by the German philosopher Gottfried Leibniz. A typical theodicy hinges on the argument that evil is the result of God letting people have free will. If mankind didn't have the choice of doing good or evil, then people wouldn't be any different from machines.[1] While theodicy raises interesting points about the nature of evil, it doesn't help us understand whether there is a science of evil. That's because, within the known laws of nature and science, evil as a supernatural force, an entity, or both, does not exist.

Evil Behavior: Genetics or Intent?

Scientific laws do not define evil with mathematical precision or experimental proof. For all we know, there are a thousand shades of evil, ranging everywhere from murder to theft to bullying to innocent comments that make someone commit suicide. If we look at evil as not being supernatural, however, but as the result of a wide range of human behaviors, both intentional and innocent, then we can examine it in a more scientific manner.

Recent explanations for the behavior of serial killers usually revolve around extra Y chromosomes or insanity. It's been suggested by scientists that a genotype of XYY causes men to fly into fits of anger, into rages, into acts of great violence. Insanity, as we all know, is often used as a defense for violent crime. Yet not only serial killers murder people. Seemingly normal people commit acts of evil every day. Could it be possible that evil is programmed into our genes, into our DNA?

In 2002, a group of scientists working for the Institute of Psychiatry in London claimed that they had discovered "the criminal gene." According to the scientists, the particular gene was strongly linked to criminal and antisocial behavior. Children from poor circumstances were nine times more likely to act unlawfully when compared to other children living in similar circumstances if they had a particular variation of the gene.[2]

Needless to say, not everyone agreed with the findings of the Institute of Psychiatry—especially when the Nuffield Council on Bioethics, located in London, declared that a criminal's genetic makeup should be taken into account during his trial and his sentencing.[3] Politicians on the right immediately suspected a plot to pardon criminals for the most unforgivable sins using the argument that such acts were a result of genetics, not intent. Politicians on the left fretted about the possible discovery of a "gay" gene, with implications that homosexual behavior could be modified or changed by gene therapy.

Rising like tidal waves from both sides of the political spectrum came dire warnings of early twentieth-century eugenic programs that championed sterilizing criminals so that their traits would be

wiped out of future generations. These stories were quickly followed with talk of Nazi genetic experiments in the 1930s, forced
sterilizations of specific groups, and the possible existence of an
"alcoholic" gene that turned ordinary Englishmen into drunkards.
No one wanted to be categorized by his or her genetic code. Nor
did either political party want criminals set free due to a mix-up in
their genetic code. At present, there's no agreement on whether the
criminal gene actually influences behavior or not. The argument
has moved out of the scientific community into politics, and it's
doubtful that any resolution to the question will be soon found. Evil
remains a matter of behavior, not genetics.

Is it evil to ruin the reputation of a coworker who leaves your
company? Is it evil to hire only people of your own race or religion,
while ignoring those of other backgrounds who have better quali-
fications? Is it evil to enter another country with troops and start a
violent war? Is it evil to let babies and the elderly starve and live on
the streets while you live in a comfortable home with plenty to eat?

As we think about the shades of evil, we realize that it doesn't
come in simple flavors, with simple explanations such as an extra
Y chromosome, insanity, and the criminal gene. It comes in many
flavors and with many attitudes. It comes in as many varieties as we
have individuals on Earth.

Fuzzy Logic Tackles Evil

If anything, evil can be analyzed much as one applies fuzzy logic to
other complex problems that require approximate reasoning. In
fuzzy logic, we view everything in shades of gray, rather than as
black and white, true and false, good and bad.

Traditional logic dictates that everything is binary—that is,
either it's true or false, one or zero, yes or no. I'm either in my office
right now, or I'm not in my office. Yet what if I'm in the open door-
way to the office? Does this mean I'm in the office or outside of it?
Clearly, it means that I'm in between—in the fuzzy area of gray.

The grass is green: true or false? Well, it's green here, but brown
there, and somewhat moldy and yellowish over there. It's between
states. Maybe it's 0.9 percent green overall. And that's in your yard.

In my yard next door, the grass may be only 0.0002 percent green overall. In fact, I may have only one blade of grass in my lawn, and it may be 0.0001 percent green. When I view your grass in the evening from my backyard window, your grass may appear 0.3 percent green, while the grass in my yard may appear 0.005 percent green. It all depends on the circumstances, the environment, and many other variables.

So why should human nature be any different? We are individuals with different minds, all of us, and the sets of variables that make up our personalities, emotions, and thought processes are seemingly limitless. We are all fuzzy sets of logic. What's evil in one context may not be evil in another. We can exhibit excellent morals by our society's standards in one environment, yet exhibit sadistic behavior in another. We can be loyal to some friends, yet extremely disloyal to friends who end up working with us. While working for a boss who condones unethical and fundamentally immoral behavior, we may knife a friend in the back. While working for a different boss, one who is highly ethical and moral, we may be terrified to knife the same friend. As the social psychologist Carol Tavris wrote, "The assumption that a moral failing in one domain reveals something profoundly important about a person's entire character, or predicts his or her behavior in other situations, is wrong."[4]

It is because our evil is driven from inside in so many shades of gray that we need our world with walls. If you do A and B and C and D, but you also don't do Q and S and T, but you sometimes do P and you often do Y, yet you always do Z after doing V and X, but you sometimes do Z after doing X and Y—if you are in a particular time and place and your mind is in a particular shade of gray on the evil plane, perhaps you will scream and curse, you may hit someone, you might say something you regret, or maybe you will do something even worse, something truly evil. The laws are in place to try and protect us from this evil. The guards are standing duty on the wall, and they're ready to shoot and kill. Just go to any international airport today and gaze at the soldiers armed with automatic weapons surveying the crowds.

The Roots of Obsession

Obsession is the root of many of Stephen King's plots. In Hitchcock's film *The Birds*, a woman is obsessed with a man in Bodega Bay, in *Vertigo* a man is obsessed with a woman who commits suicide, and in King's *Carrie*, teenagers are obsessed with bullying an outcast. Do supernatural forces control our obsessions, or, like evil, are obsessions created by our own minds? By now, you've probably guessed that we're going to argue in favor of obsessions that have some sort of scientific foundation. First, let's define the term and provide some recent, rather infamous, examples.

Obsession is defined as a compulsive preoccupation with a fixed idea or feeling. Great anxiety and stress usually accompany obsessions. People tend to become irrational, driven by their obsessions, unable to damper or stop them.

John W. Hinckley Jr.

A classic case of obsession involves the attempted assassination of U.S. president Ronald Reagan in 1981. On March 30, Reagan and his entourage left the Washington Hilton Hotel, where John W. Hinckley Jr. tried to shoot him. Instead, Hinckley shot James Brady, Reagan's press secretary, in the head and wounded two other people. With many eye witnesses—indeed, with the attack filmed on camera—Hinckley was immediately arrested and admitted his guilt.

Despite the evidence and his confession, however, Hinckley's defense claimed that he was innocent by reason of insanity. The important part of this story—at least to us in this chapter—is that Hinckley claimed he was insane because for years he'd been obsessed with the movie star Jodie Foster.

Extensive court-appointed psychological examinations concluded that Hinckley was sane when he fired the shots because he had planned the assassination for a long time. The defense, on the other hand, provided a host of reasons for his insanity, and ultimately, the jury agreed that he was indeed insane and not responsible for his evil actions. Hinckley was acquitted.

Hinckley's obsession with Jodie Foster began when he saw her debut in the movie *Taxi Driver*, in which she played the role of a prostitute. A psychotic taxi driver named Travis Bickle saves Foster from her pimp. Hinckley felt that he was Travis Bickle, and he started dressing like the character, talking like him, and using guns. A year later, Hinckley turned to target practice, while he contemplated suicide and wrote to and called Foster repeatedly, claiming to love her.

When Foster refused to yield to her obsessed stalker, Hinckley took increased action to get her attention. His idea was crazy: he would kill someone famous to get Foster to pay attention to and love him. At first, he decided to kill President Jimmy Carter. He stalked Carter throughout 1980. He couldn't go through with the Carter assassination and instead contemplated suicide at the location where, weeks earlier, another obsessed fan, Mark David Chapman, had murdered John Lennon. Finally, he ended up firing shots at Reagan, and due to his insane obsession with Jodie Foster, he was deemed not guilty. He did spend many years in a mental institution to pay for his crimes.

Mark David Chapman

On December 8, 1980, John Lennon was killed by Mark David Chapman in front of Lennon's New York City apartment, the Dakota. There were numerous witnesses to the crime, and after killing Lennon, Chapman told the doorman, "I just shot John Lennon." As with Hinckley, Chapman's crime had witnesses and a confession.

Chapman had a history of mental illness and, after reading J. D. Salinger's *The Catcher in the Rye*, expressed a great desire to be like Holden Caulfield in the novel. He became obsessed with John Lennon and even married a Japanese American woman who, most likely, reminded him of Yoko Ono. Chapman felt he was nothing, a nobody, and within the confines of his depressing life, he decided that Lennon was a lousy hero, a phony, as Holden Caulfield might say. He tried to murder Lennon twice, he testified in court, but didn't have the courage to pull the trigger. His third attempt succeeded. In Chapman's case, he pleaded guilty and was sentenced to a prison term of twenty years to life.

Fuzzy Obsessions

As with evil, obsessions also come on the fuzzy scale, somewhere in the gray between black and white. Someone with an obsession for sushi or apple pie isn't as dangerous as someone with an obsession with a celebrity. If we view our brains as individual minds, each different from the other, the fuzzy scale applies to all human emotions, behaviors, and conduct. If we understand how our brains work, then by extension we understand why some of us are more evil than others, and why some of us become psychotically obsessed with Jodie Foster or John Lennon, with a man in a small town, with the wife of a friend, or with It.

Roots of Obsession: Neurological or Psychological?

Evolutionary psychologists tell us that our minds evolved long ago to help us survive. We learned how to recognize one another's faces, how to recognize and cope with cheating, how to choose mates, and how to talk to one another. Various groups of neurons might handle something like language and be located in one area of the brain—in the case of language, in what is known as Broca's area. Other groups of neurons might not be located in the same area. In this way of looking at the brain, small modules of neurons feed information to larger modules. Even smaller modules feed information to the small modules, and so forth, until, at the lowest level, an individual neuron fires during specific events.

A neuron fires electrochemically, meaning that chemicals produce electric signals. When chemicals in our bodies have an electric charge, they are termed ions, and ions in the nervous system include sodium and potassium, each with one positive charge; calcium with two positive charges; and choride with one negative charge. Neurons are surrounded by a semipermeable membrane that lets some ions pass through while blocking other ions.

When the neuron is not firing, the inside of the cell is negative compared to everything immediately surrounding the cell. The ions keep trying to pass from the inside of the neuron to the outside and vice versa, with the membrane controlling the balance. For example, when the neuron is not firing, potassium ions pass easily through the membrane, and for every two potassium ions the membrane allows to enter the neuron, it allows three sodium ions to leave. Basically, there are more sodium ions outside the neuron and more potassium ions inside it. At rest, when the neuron is not firing, the difference in voltage between the inside and the outside of the neuron is approximately −70 millivolts, meaning that the inside of the neuron is 70 millivolts less than the outside.

When a neuron sends information down an axon away from the cell body, neuroscientists say that there is an action potential or that a spike has occurred. The action potential is created by a depolarizing current, which creates electrical activity. An event, or stimulus, occurs that moves the resting potential of −70 millivolts toward 0 millivolts. The stimulus causes the sodium channels to open in the neuron, and because there are more sodium ions outside the neuron than inside, sodium ions flood into the neuron. Because sodium ions have a positive charge, the neuron becomes more positive and depolarized. When depolarization shifts downward to approximately −55 millivolts, the neuron fires an action potential, which is known as the threshold. If the neuron never reaches its threshold, it won't fire.

The potassium channels open after the sodium channels, and when they do, potassium moves out of the cell, reversing the depolarization. The sodium channels start closing, and the action potential reverses, moving back toward −70 millivolts.

So what does this have to do with evil tendencies and obsessions? It explains what's happening inside the brain to cause experiences, behaviors, actions, thoughts—and obsessions. In 2003, *Scientific American* reported that evidence indicates that when people think they are seeing aliens, ghosts, and demons, or when they think they are floating on the ceiling, what's really happening is a firing of neurons inside their brains, imposing upon them the fiction that they're seeing things or floating. It's all induced inside the body—specifically, by the neuronal connections in the brain.[5]

There's plenty of scientific evidence to support this notion. For example, the neuroscientist Olaf Blanke provokes out-of-body experiences in people by stimulating the right angular gyrus in the temporal lobe.[6] And the neuroscientist Michael Persinger subjects patterns of magnetic fields to patients' temporal lobes to induce all sorts of supernatural and out-of-body experiences. He forces the neuron firing patterns to become abnormal and unstable, with the result that patients experience abnormal psychological states. With six hundred patients studied now, he said that these abnormal and unstable neuronal events could occur naturally during times of great stress, when we fast, when we fly at high altitudes, and when our blood sugar changes dramatically.[7]

It's quite possible, therefore, that Hinckley's obsession was initially caused by unstable neuronal firing patterns triggered by a traumatic event. In Chapman's case, it may have been caused by the trauma of being in a mental institution and/or consuming drugs. In both cases, and in the cases of Stephen King's characters, the neuronal stimuli causing obsessions may be due to depression, poor nutrition, a near-death experience, a change in blood sugar, or fasting.

At present, scientists studying obsession and obsessive-compulsive disorder (OCD) are broken into two groups feuding over the cause of the malady. One group believes that OCD is purely a psychological disorder. People in this group think that obsessive-compulsive behavior is caused by individuals feeling that they are personally responsible for their obsessive thoughts. This overblown sense of responsibility makes the victims more anxious and keeps the obsessive thoughts in their minds. On the other side of the argument are scientists who believe that OCD is the result of abnormalities of the brain. Most scientists believe this argument.[8]

One of the most popular theories explaining OCD behavior involves an imbalance in the brain chemical serotonin. Serotonin and other chemicals known as neurotransmitters travel from nerve cell to nerve cell across synapses. Serotonin is believed to regulate memory, anxiety, and sleep. When serotonin is released by one cell, it enters another cell through a special area in the cell known as the receptor. In people with OCD, however, some receptors are believed to block serotonin from entering a cell. That brings about a deficiency of neurotransmitters in important areas of the brain, which results in obsessive-compulsive behavior.[9]

Physicians are experimenting with chemicals known as selective serotonin reuptake inhibitors. These drugs prevent nerve cells that have just released serotonin from reabsorbing the chemical. Thus, the serotonin is available for other cells. Scientists hope this treatment will result in a cure for OCD and will make obsessive behavior a thing of the past. This will force Stephen King to search for a new and different plot device to motivate his villains.

The Seeds of Fear

Fear is typically thought of as a feeling of anxiety caused by the possible or real presence of danger. As with any emotion, such as obsession, jealousy, and love, fear comes in many shades of gray. Mild fear might refer to a general feeling of uneasiness or apprehension. An example is the fear we feel if people make unkind comments that result in our worrying about their loyalty and friendship. Severe fear, of course, refers to the terror a person feels if someone is breaking into his house or chasing him at night through a forest.

When we experience fear, our bodies react in certain ways. Our heart rates increase, sending blood pumping through our veins, as our bodies make sure that our muscles have enough oxygen to face danger. Our blood pressure rises as we grow fearful. In a stress situation, our bodies shut down any unnecessary systems, including our digestive systems. We therefore produce less saliva, a digestive fluid, and our mouths turn dry.

When we are scared, certain primitive traits prepare us for action. Blood vessels near the skin tighten so as to reduce bleeding if we're injured. The pupils of our eyes dilate to focus on any movement. The hairs on our skin stand up, making us more sensitive to movement. The more scared we are, the more noticeable the symptoms.[10]

King excels at provoking fear in readers. In 1981's *Danse Macabre*, he explains that his writing process is a "dance" in which he exposes the private fears of his readers. What could be more horrifying than having a member of your own family trying to kill you, as portrayed in *The Shining*? For any working writer, King's *Misery* is a masterpiece of fear, as an insane killer nurse is so obsessed with a writer's work that she enslaves and tortures him to make him write what she wants. The nurse may be the physical manifestation of the writer's muse, which forces him to continue writing material that he's long since grown bored with. In King's short stories, fear also dominates, as in "Night Surf," where six young people survive a deadly flu virus, then huddle on a Maine beach, listening to the radio and pondering their imminent deaths.

Neuroscientists believe that the seeds of fear lie in a part of the brain called the amygdala, which is derived from the Greek *amygdala*, meaning "almond." Each amygdala is an almond-shaped group of neurons located in the medial temporal lobes in humans and other complex vertebrates. The temporal lobes are in the cerebrum at the sides of the brain. The amygdalae encompass nuclei such as the cortical nucleus, the centromedial nucleus, and the basolateral complex, which can be divided further into basal, lateral, and accessory basal nuclei.

In neuroanatomy, a nucleus refers to a structure composed primarily of gray matter that acts as a transit point for electrical signals in one neural subsystem. Gray matter consists of unmyelinated neurons, where myelin is a phospholipid layer of electrical insulation surrounding the axons of neurons. The myelin layer, or sheath, enables impulses to propagate more quickly, almost "hopping" down the fibers, whereas in unmyelinated fibers, impulses move as waves. So the gray matter is made of nerve cell bodies and short axons and dendrites that do not have myelin sheaths. It basically processes information from sensory organs and motor stimuli.

The amygdalae perform vital roles in our memories of emotional reactions and in our processing of fear and aggression. They are part of the limbic system, which is the portion of the brain that integrates our emotional states with our memories of physical sensations, that influences our motivation to do things, and that instills fear in us.

When we are frightened, the amygdalae transmit impulses to the hypothalamus, which links the nervous system to the endocrine system through the pituitary gland. The hypothalamus regulates various metabolic and autonomic processes, such as body temperature, thirst, and hunger. When the hypothalamus receives impulses from the amygdalae, it activates the sympathetic nervous system, which in turn triggers the sympatho-adrenal response, commonly known as the fight-or-flight response. Acetylcholine is secreted, along with adrenaline. As mentioned earlier, our heartbeats may increase, and blood vessels may become constricted. Our pupils may dilate, we may sweat, and our blood pressure may rise.

Along with transmissions to the hypothalamus, the amygdalae also send impulses to the reticular nucleus for increased reflexes. The reticular nucleus transmits signals to the nuclei of the facial and trigeminal nerves so our faces display fear. And along with all of these sensations, our emotions rise and flutter. In short, we are in a state of terror.

When we are frightened, sensory input enters the amygdalae and forms associations with memories of being frightened, of these specific types of sensory input. The association between frightening events and the sensory stimuli may be directly affected by the potential of the involved synapses to react quickly due to what's known as long-term potentiation.[11]

Long-term potentiation refers to the prolonged enhancement of the efficiency of the synapses between neurons. Scientists believe that long-term potentiation contributes to synaptic plasticity, or the ability of the synapses between neurons to change in strength over time. It is a foundation of learning and memory.

If you encounter It once, you will always be afraid of It. The terrifying first encounter remains in your mind, stored in the amygdalae. When It reappears, the amygdalae remembers the horrifying experience and releases two hormones, adrenaline (also called epinephrine) and norepinephrine, into your bloodstream. These hormones make your heart beat faster and supply blood to your muscles so you can run quickly from Its presence.

The Columbia University scientist and 2000 Nobel Prize–winner Eric Kandel has discovered two genes that can be used to inhibit the amygdalae from learning how to fear things. In May 2006, a summary of his work at the Howard Hughes Medical Institute stated:

> Fear in mice, monkeys, and people is mediated by the amygdala, a structure that lies deep within the cerebral cortex. To develop a molecular approach to learned fear in the mouse, we identified two genes as being highly expressed both in the lateral nucleus of the amygdala—the nucleus where associations for Pavlovian learned fear are formed—and in the regions that convey fearful auditory information to the lateral nucleus.[12]

One gene is GRP, and the other is stathmin, an inhibitor of microtubule formation that is highly expressed in the amygdalae. Deficits in either the GRP receptor or stathmin cause mice to be more aggressive, more bold, and acking in fear.

Studies involving humans and the GRP gene are still in the future. Yet scientists are already discussing the possibility of fear serum that would be given to children to banish all fears and phobias from their minds. Whether such youngsters would grow up to be superior or inferior to normal humans is a matter for much debate. Fear is an important defense mechanism in all animals. A man or a woman without fear might be the next step forward in human evolution—or the next step backward.

It's probably best to allow certain fears to exist, such as the fear of killer dogs or psychotic stalkers. If we take away fear, then how will we combat It or the Tommyknockers?

NOTES

1: From Proms to Cells: The Psychic World of Stephen King

1   Daniel Dunglas Home, James Randi Educational Foundation, www
    .randi.org/encyclopedia/Home,%20Daniel%20Dunglas.html
    (accessed November 11, 2006).

2   Tim Underwood and Chuck Miller, *Feast of Fear: Conversations with
    Stephen King* (New York: Warner Books, 1993), 105.

3   James Longrigg, *Greek Rational Medicine: Philosophy and Medicine from
    Alchaeon to the Alexandrians* (New York: Routledge, 1993), 58.

4   Andrew Cunningham, *The Anatomical Renaissance: The Resurrection
    of the Anatomical Projects of the Ancients* (Aldershot, U.K.: Ashgate,
    1997), 12.

5   Lois H. Gresh, *Exploring Philip Pullman's His Dark Materials: Dust,
    Angels, Souls, and Weird Science* (New York: St. Martin's Press,
    expected publication November 2007). The material in this section is
    based on ideas in Lois's earlier book, written in 2005 and awaiting
    publication at the time of this writing.

6   Ibid.

7   *New Catholic Encyclopedia* (New York: McGraw-Hill, 1967), 460.

8   Originally by Gilbert Ryle, *The Concept of Mind* (New York:
    Hutchinson, 1949), though the phrase "ghost in the machine" has
    become a common term used in many texts and articles.

9   Michio Kaku and Jennifer Thompson, *Beyond Einstein: The Cosmic Quest
    for the Theory of the Universe* (New York: Anchor Books, 1995), 48.

10   Amit Goswami, Richard E. Reed, and Maggie Goswami, *The Self-Aware Universe: How Consciousness Creates the Material World* (New York: Penguin Putnam, 1995), 4.
11   Ibid., 6–7.
12   Gresh, *Exploring Philip Pullman's His Dark Materials*, ibid.
13   Hellmut Wilhelm and Richard Wilhelm, *Understanding the I Ching: The Wilhelm Lectures on the Book of Changes* (Princeton, N.J.: Princeton University Press, 1995), 5. According to Princeton University Press, Richard Wilhelm was the West's foremost translator of the I Ching.

2: On the Highway with Stephen King

1   Stephen King, "Trucks," in *Night Shift* (New York: Doubleday, 1978), 139.
2   Kim Garnel, "Morse Code Is Entering 21st Century," *Chicago Sun-Times*, March 1, 2006.
3   King, "Trucks," 134.
4   Jack Williamson, *Wonder's Child: My Life in Science Fiction* (New York: Bluejay, 1984).
5   The authors have written an entire book on the computers of Star Trek, with the catchy title *The Computers of Star Trek*.
6   Brian Handwerk, "'Intelligent' Cars 'Talk' with Highway, One Another," *National Geographic News*, May 21, 2004, http://news .nationalgeographic.com/news/2004/05/0521_040521_smartcars .html/ (accessed November 11, 2006).
7   Ibid.
8   Ibid.
9   Simon Taylor, "Intelligent Cars' Initiative Could Save Lives: European Commission Unveils Latest Digital Safety Devices for Cars," IDG News Service, February 21, 2006, www.infoworld.com/ article/06/02/21/75550_intelligentcars_1.html/ (accessed November 11, 2006).
10   Ibid.
11   "What Is Friendly AI?" Singularity Institute for Artificial Intelligence, www.singinst.org/friendly/whatis-print.html/ (accessed November 11, 2006).
12   Ibid.

3: They Came from Outer Space

1   Seth Shotek, *Sharing the Universe* (Berkeley Hills, CA: Berkeley Hills Books, 1998), foreward by Frank Drake, i–ii.
2   David Darling, *Life Everywhere: The Maverick Science of Astrobiology* (New York: Basic Books, 2001), xiii.
3   Ibid., xi.
4   Interview with Michio Kaku in *Astrobiology Magazine*, April 26, 2004, as reported at www.astrobio.net/news/ (accessed November 11, 2006).
5   Ibid.
6   Stefan Lovgren, "Flying Whales, Other Aliens Theorized by Scientists," *National Geographic*, http://news.nationalgeographic.com/news/2005/05/0520_050520_tv_aliens.html/ (accessed November 11, 2006).
7   Ibid.
8   Rod Serling, "The Monsters Are Due on Maple Street," *Twilight Zone*, March 4, 1960.

4: The Fourth Horseman

1   Stephen King, *The Stand, Complete and Uncut* (New York: Doubleday, 1990), 70
2   Ibid., 1,149.
3   William C. Patrick III, "The Threat of Biological Warfare," *Washington Roundtable on Science and Public Policy*, February 13, 2001.
4   "Small World Phenomena," *Wikipedia*, http://en.wikipedia.org/wiki/Small_world_phenomena (accessed November 11, 2006).
5   "The Black Death," *Wikipedia*, http://en.wikipedia.org/wiki/Black_Death/ (accessed November 11, 2006).
6   "Plague: Fact and Information," the Preparedness Center, Preparedness Industries Inc., www.preparedness.com/plagfacin.html (accessed November 11, 2006).
7   Ibid.
8   "The American Experience, 1918 Influenza Timeline," www.pbs.org/wgbh/amex/influenza/timeline/index.html/ (accessed November 11, 2006).

9   "Spanish Flu," *Wikipedia*, http://en.wikipedia.org/wiki/Spanish_flu/ (accessed March 11, 2006).

10   Mike Adams, "Bird Flu Timeline: A History of Influenza from 412 B.C.–A.D. 2006," http://www.newstarget.com/017503.html/ (accessed February 6, 2006).

11   Ibid.

12   Ibid.

13   "Avian Flu Vaccine to Have 6-Month Lag," *USA Today*, March 14, 2006.

14   James Kirkup, "Bird Flu: The Secret Cabinet Document," Scotsman .com, April 3, 2006 (accessed November 11, 2006).

15   WHO bulletin 47 (1972): 259

16   "AIDS Created as Biowarfare, Says Nobel Laureate," *Conspiracy Planet*, September 23, 2006, www.conspiracyplanet.com/channel .cfm?channelid=34&contentid=1595/ (accessed November 11, 2006).

5: Up the Dimensions with Stephen King

1   *Insomnia* (New York: Viking Books,1994).

2   Robert Hatch, "Sir Isaac Newton," *Encyclopedia Americana* (1998): 288–292, web.clas.ufl.edu/users/rhatch/pages/01-Courses/current-courses/08sr-newton.htm/ (accessed November 11, 2006).

3   Tensor equations that require differential geometry for their solutions are beyond the scope of this book.

4   "History, Quarks," *Wikipedia*, http://en.wikipedia.org/wiki/ Quark#Current_quark_mass/ (accessed November 11, 2006).

5   Dennis Overby. "String Theory, at 20, Explains It All (or Not)," *New York Times*: Science, December 7, 2004.

6   Brian Greene, *The Elegant Universe* (New York: Vintage, 1999, 2003), 138.

7   Ibid., 141.

8   Overby, "String theory."

9   Ibid.

10   Ibid.

11.   Ibid.

6: Traveling in Time with Stephen King

1   Stephen King, "The Langoliers," in *Four Past Midnight*, (New York: Viking, 1990), 127.

2   Ibid.

3   Ibid., 198.
4   "Time Travel," *Wikipedia*, http://en.wikipedia.org/wiki/Time_travel/ (accessed February 26, 2006).
5   Paul Rincon, "Wormhole 'No Use' for Time Travel," BBC NEWS, May 23, 2005 (accessed November 11, 2006).

7: Parallel Worlds

1   Stephen King, *From a Buick 8* (New York: Scribner, 2002), 322.
2   Ibid., 325.
3   Lois H. Gresh and Robert Weinberg, *The Science of Supervillains* (New York: John Wiley & Sons, 2004), 143–57.
4   "Parallel   Universes,"   www.bbc.co.uk/science/horizon/2001/paralleluni.shtml/ (accessed November 11, 2006).
5   Max Tegmark, "Parallel Universes," www.sciam.com, 2003, page 41.
6   Ibid.
7   Ibid.
8   Marcus Chown, "It Came from Another Dimension," *New Scientist* (December 2004): 31.
9   "The Truth about Warp Drive," *New Scientist* (March 16, 2002): 26.
10   Michio Kaku, *Parallel Worlds* (New York: Doubleday, January 2005), 5.
11   Ibid.
12   Ibid.
13   For more information, see Professor Guth's fascinating Web site at http://web.mit.edu/physics/facultyandstaff/faculty/alan_guth.html/ (accessed November 11, 2006).
14   Kaku, 88.
15   Ibid., 88.
16   Ibid., 14.
17   Ibid., 93.

8: Longevity and Genetic Research

1   Dr. Alexander Leaf, "Long-Lived Populations (Extreme Old Age)," in *Principles of Geriatric Medicine and Gerontology*, 2nd ed., edited by W. R. Hazzard, R. Andres, E. L. Bierman, and J. P. Glass (New York: McGraw-Hill, 1990).
2   Michael G. Zey, *The Future Factor: The Five Forces Transforming Our Lives and Shaping Human Destiny* (New York: McGraw-Hill, 2000),

51. Zey is the executive director of the Expansionary Institute, a consultant to Fortune 500 companies and government agencies.

3   Natalie Angier, "Surprising Role Found for Breast Cancer Gene," *New York Times*, March 5, 1996.

4   J. Medeleine Nash, "The Immortality Enzyme," *Time* (September 1, 1997): 65.

5   Clare Thompson, "On the Horizon," *Time* (May 12, 2001), www .time.com/time/magazine/printout/0,8816,17690,00.html/ (accessed November 11, 2006).

6   S. Jay Olshansky and Bruce A. Carnes, *The Quest for Immortality: Science at the Frontiers of Aging* (New York: W. W. Norton, 2001), 187.

7   Thompson, "On the Horizon."

8   Zey, *The Future Factor*, 57.

9   Ibid., 59.

10   B. N. Ames, M. K. Shigenaga, and T. M. Hagen, "Oxidants, Antioxidants, and the Degenerative Diseases of Aging," Proceedings of the National Academy of Sciences of the United States of America,www.ncbi.nlm.nih.gov/entrez/query.fcgi?cmd=Retrieve&db =PubMed&dopt=Abstract&list_uids=8367443/ (accessed November 11, 2006).

11   Jeff Goldberg, "Artificial Heart," *Life* (Fall 1998), 85.

12   John M. Leferovich, et al., "Heart Regeneration in Adult MRL Mice," *Proceedings of the National Academy of Sciences of the United States of the America* (August 14, 2001), www.pnas.org/cgi/content/full/98/17/ 9830/ (accessed November 11, 2006).

13   "Artificial Skin," *Popular Science* (September 1997): 15.

14   "Tech 2010," *New York Times Magazine*, June 11, 2000.

15   Frederick B. Rudolph and Larry V. McIntire, ed., *Biotechnology: Science, Engineering, and Ethical Challenges for the Twenty-First Century* (Washington, D.C.: Joseph Henry Press, 1996), 88–89.

16   K. Eric Drexler, *Engines of Creation: The Coming Era of Nanotechnology* (New York: Anchor Books/Doubleday, 1986), 124. If you haven't read this book, you must do so immediately. It's one of the most significant books published during the last two decades.

9: Evil, Obsession, and Fear

1   "Theodicy," *Wikipedia*, http://en.wikipedia.org/wiki/Theodicy/ (accessed November 11, 2006).

2   "British Scientists Discover Criminal Gene," *ABC Science Tech*, May 8, 2002, http://abc.net.au/news/scitech/2002/08/item2002080222 5123_1.htm/ (accessed November 11, 2006).

3   "Criminal Gene Could Mean Lighter Sentencing," *GKT Gazette*, October 2002, www.gktgazette.com/2002/oct/news.asp#6/ (accessed November 11, 2006).

4   Carol Tavris, "All Bad or All Good? Neither," *Los Angeles Times*, September 20, 1998, M5.

5   Michael Shermer, "Demon-Haunted Brain," *Scientific American* (March 2003): 32

6   Olaf S. Blake, T. Ortigue, T. Landis, and M. Seeck, "Neuropsychology: Stimulating Illusory Own-Body Perceptions," *Nature* 419 (September 19, 2002): 269–70.

7   Michael A. Persinger, *Neuropsychological Bases of God Beliefs* (New York: Praeger, 1987), and Persinger, "Paranormal and Religious Beliefs May Be Mediated Differently by Subcortical and Cortical Phenomenological Processes of the Temporal (Limbic) Lobes," *Perceptual and Motor Skills* 76 (1993): 247–51.

8   "Causes of OCD," *BBC—Science & Nature*, www.bbc.co.uk/science/humanbody/mind/articles/disorders/causesofocd.shtml/ (accessed November 11, 2006).

9   Ibid.

10 "Fear Gets Our Heart Racing," RTÉ 2006, www.rte.ie/tv/scope/SCOPE3_show01_fear.html/ (accessed November 11, 2006).

11 Kerry Ressler and Michael Davis, "The Amygdala Is the Primary Brain Region Involved in Fear-Conditioned Learning," *Journal of the American Academy of Child and Adolescent Psychiatry* 42:5, (May 2003): 612–15.

12 "Cell and Molecular Biological Studies of Memory Storage," research abstract, Howard Hughes Medical Institute, www.hhmi.org/research/investigators/kandel.html/ (accessed November 11, 2006).

INDEX

Printed in the USA
CPSIA information can be obtained
at www.ICGtesting.com
JSHW021320221024
72173JS00001B/15